Grounding Grounded Theory

GUIDELINES FOR QUALITATIVE INQUIRY

Ian Dey
Department of Social Policy and Social Work
University of Edinburgh
Edinburgh, United Kingdom

ACADEMIC PRESS
San Diego London Boston New York Sydney Tokyo Toronto

This book is printed on acid-free paper. ∞

Academic Press
a division of Harcourt Brace & Company
525 B Street, Suite 1900, San Diego, California 92101-4495, USA
http://www.apnet.com

Academic Press
24-28 Oval Road, London NW1 7DX, UK
http://www.hbuk.co.uk/ap/

Library of Congress Catalog Card Number: 98-83122

International Standard Book Number: 0-12-214640-9

PRINTED IN THE UNITED STATES OF AMERICA
99 00 01 02 03 04 EB 9 8 7 6 5 4 3 2 1

To Diana, who has kept me well grounded throughout this project.

CONTENTS

3 Categories

4 Categorization

5 Coding

6 Concluding

PROLOGUE

A few years ago, at an international conference in Breckenridge, Colorado, I presented some software for analyzing qualitative data. To my surprise, most of my audience assumed that the software I presented was based on a grounded theory approach to analyzing data.

I was intrigued—and, I admit, flattered—by the kudos that seemed to accompany an association with grounded theory. In a social science context it is hard to resist the claim that one's work has theoretical import—and theory which is "grounded" at that! Participants at the conference seemed to attach a certain prestige to software that could support the development of grounded theory. Some developers of other software applications quite explicitly made reference to grounded theory as their primary source of inspiration. Indeed, it seemed that grounded theory and software applications for qualitative analysis were to some extent bound together, with each gaining legitimation from the other. While I was flattered by the association with grounded theory, I found this rather a puzzle, for I had designed the software without any reference to grounded theory.

This book owes its origin to that initial puzzle. Although it began as an exploration of the relationship between software applications and grounded theory, it rapidly transformed into an inquiry into grounded theory in its own right. While I have set out to satisfy my own curiosity about grounded theory, I hope this book, if not providing answers, will at least raise some relevant questions on which others may find it useful to reflect.

Unusually (I imagine) for a book on methodology, this book begins from a position more of ignorance than of expertise—for I make no claims to be an authority on grounded theory. In this regard, I suspect I am closer to those of my contemporaries who, through the use of software for qualitative analysis, are encountering grounded theory for the first time—and wondering what to do with it. With the proliferation of software applications, no doubt others too may be puzzling over the connections claimed between grounded theory and software for qualitative analysis. If so, I share their concerns and hope this book will help to clarify them.

But this is not a "how to" text. I do set out to consider what grounded theory has to offer to qualitative inquiry, but my intention is not to prescribe methods but to reflect on a methodology. I am therefore less concerned with detailed prescriptions of how to use grounded theory than with assessing the kind of account of qualitative research that grounded theory offers. What guidelines does it offer for qualitative inquiry?

In such an enterprise, it is conventional to identify one's values (insofar as one is able) at the outset. To this endeavor, I no doubt bring some biases arising from my own background and experience. For example, as a former son of the manse, my Presbyterian upbringing has bestowed an abiding suspicion of received authority and scholastic dogma. Although I am no longer Presbyterian, I still find appealing the idea that individuals have to make their own peace with their gods (or demons). So I start with an initial skepticism regarding authoritative claims associated with any schools of thought. With regard to social research, this skepticism would leave to each of us the task of assessing, selecting, and reflecting upon the methods and the methodologies we wish to employ.

My outlook on methodology has been further influenced by two very different texts. One is Andreski's (1974) bitter but stimulating (if now rather dated) critique of "social science as sorcery," which castigated social scientists for their habit of constructing careers through new jargon rather than through new concepts—a case of old wine continually being recast in new bottles. Andreski raised and reinforced my suspicions concerning the legitimating rather than the analytic functions of claims to knowledge in social science. Another influence was a critique by Hindess (1977) of the major analytic perspectives of interpretive and positivist sociology. Hindess tried to demonstrate that these perspectives were internally inconsistent, in each case preaching epistemological principles and methodological canons that bore little relationship to social science as practiced—even by the preachers.

Perhaps because of this skeptical bias, I find the idea that social scientists preach one thing while practicing another irresistibly attractive. That too probably stems from my early upbringing, notably my adolescent appreciation that life in the church in general and in the manse in particular failed (inevitably, of course) to conform to professed Christian principles (or pretensions). Nevertheless, I think this provides a useful starting point for any appreciation of methodological issues. It encourages us to question established authority. It fosters that critical attitude to received wisdom, which, for the physicist Richard Feynman, was the hallmark of a scientific approach (Gleick, 1994). Feynman always had to work things (and theories) out for himself. He believed in the accumulated wisdom of science, but he never took it on trust.

Previously established theories had to be grounded through his own grasp of what they involved and how they worked.

The title of this book refers to "grounding" grounded theory. The Concise Oxford Dictionary (which supplies the various definitions used in this text) offers two rather contradictory senses of the verb "to ground." One is to "run ashore" or "to strand," while the other is "to base" or "to establish." One sense is positive—that ideas can be established on the basis of some fact or authority or, in this case, some critical reflection. The other sense is rather negative—that ideas that are floating without foundation should be run ashore as one might ground children who have abused their freedom by staying out too late. While we may rightly want to ground (in the negative sense) the use of methodology as legitimation, this book is an effort to ground grounded theory in the positive sense—to be critical but also constructive in reflecting on the methodological principles that grounded theory has to offer. This effort requires a review of the ways in which grounded theory approaches some of the key issues in qualititative analysis, such as coding and categorization, the analysis of process, and the generation of theory.

Such are my ambitions. My efforts to realize them are directed almost entirely to the principles of grounded theory rather than to its practice. This might be regarded as a grave drawback in any methodological text, especially given the conviction (which I share) that researchers seldom practice what they preach. However, this disparity between principle and practice may also justify my approach, since the way those using grounded theory actually go about their business may bear little relation to the principles that they themselves profess. Nevertheless, we have to give some account of what we do; and, indeed, the more reflexive we are about our research practice, the better. However, we should also treat claims that we know what we are doing with some skepticism. I am bemused, for example, by the absurdity of a recent text which, on the one hand, argues that we need to recognize "a text's desire to assert its own power over the reader" (Lincoln & Denzin, 1995, p.579), yet, on the other hand, persists with labels such as "positivist" and "postpositivist"—as though these were neutral and not denigrating terms!

This is why I shall disappoint the reader who searches these pages for a long and detailed account of the rival epistemological or ontological positions or perspectives adopted by various researchers. My sympathies in this respect are with an apposite remark made by Lincoln and Denzin in their review of some of these debates:

> These criticisms and exchanges can operate at a level of abstraction that does little to help the people who just go out and do research. (Lincoln & Denzin, 1995, p. 577)

I am particularly suspicious of claims that we must subscribe to this or that epistemological or ontological position before we can even begin to do research. The same claim is sometimes made with regard to grounded theory, whenever it is insisted that certain ontological or epistemological principles must be taken on board for the research to qualify as a bona fide example of the method.

The reader informed about these ontological and epistemological debates will recognize that I draw inspiration in this book from the realist approach to the problems of knowledge. The authors upon whom I draw most extensively—Lakoff, Archer, and Sayer—all locate themselves within a broadly realist tradition. This is not inappropriate, as grounded theory itself has some affinities with this tradition. However, I have preferred to examine the issues that these authors raise in particular methodological contexts—such as the use of categorization, the identification of causality, or the analysis of process—rather than to foreclose discussion by enunciating some abstract set of realist epistemological or ontological principles.

Such principles are rarely adequate to the purpose. Take one of the stock principles that is almost invariably argued to be a defining characteristic of qualitative research: the need to engage with a double hermeneutic and take account not only of what people do but of the particular meanings they invest in their actions and the interpretations they make of them. This tends to assume (1) that other forms of research cannot access meanings and interpretations and (2) that qualitative research must do so. The first assumption simply (and unreasonably) discounts the contribution of other methods, such as the survey or the structured interview, in investigating perceptions. The second assumption takes no account of the stratified nature of our experience as physical beings in a socially structured world, not just as persons with particular projects and perceptions. We can understand a great deal of social activity perfectly well and do so on a daily basis (in buses, shops, banks, and restaurants) without any access to the personal meanings our interactions may have for others. So even this hallowed principle, often repeated as a sine qua non of qualitative research, is open to challenge. Indeed, it must remain very doubtful how successfully qualitative researchers—whatever they profess—do or can access the personal meanings of those they study, not least because we have such difficulties in knowing our own minds. Though it is probably heresy to challenge these assumptions, especially the second, perhaps a little heresy can be refreshing if it can broach assumptions that orthodoxy simply cannot begin to question.

If it is difficult to account for what we do as researchers, it seems rather foolish to insist upon deducing research methods from a set of abstract epistemological principles that may bear little relation to practice. It is as though only once we have climbed an epistemological Mount Everest will we be permitted to play a little in the foothills of the research enterprise. I am not defending here those who "just go out and do research," preferring in the process to ignore that towering epistemological mass in favor of scrutinizing the minutest details of their progress through their data. On the contrary, I do think we should recognize and reflect upon the problems of producing knowledge through our research endeavors. But it is one thing to recognize these as issues and problems, requiring reflection, and quite another to announce that they have been solved—and, solved, moreover, in a way which claims to invalidate much of the activity that others think of as genuine research.

Hence the approach I have taken in this book, which has been to focus on some key areas of inquiry—the use of categories, the coding of data, the analysis of process, the role of theory, and the problems of validation. In each of these areas, I have tried to identify the aims and procedures of grounded theory, in either its original or later versions, though without worrying unduly about the differences between them. If Glaser and Strauss cannot agree on a definitive version, it would be impertinent (and indeed, impossible) for an outsider to arbitrate. My aim has not been a textual criticism, nor production of yet another version of grounded theory, but critical consideration (from the perspective of an interested observer) of the guidelines it offers to qualitative researchers.

ACKNOWLEDGMENTS

This book is a very personal product, and whatever follies and faults are found in it are entirely my own. It is also a collaborative project, and I acknowledge, with much gratitude, the contributions of those who made it possible—of Academic Press, who kindly invited me to write it; of my colleagues, who generously gave me time to do so; of my reviewers, whose criticisms helped me to improve (I hope) upon earlier drafts; of my editor at Academic Press, Scott Bentley, whose patience and persistence were vital in seeing the project through; and of my family, who gave me their unstinting support throughout.

CHAPTER 1

Introduction

Grounded methodology and methods (procedures) are now among the most influential and widely used modes of carrying out qualitative research when generating theory is the researcher's principal aim. (Strauss & Corbin 1997)

Let us begin this inquiry with the obvious question: what is grounded theory?

We can start with a summary of some of the main tenets of grounded theory identified in a recent text (Creswell, 1998) on five traditions in qualitative inquiry. Among other points, Creswell suggests that

- The aim of grounded theory is to generate or discover a theory
- The researcher has to set aside theoretical ideas to allow a "substantive" theory to emerge
- Theory focuses on how individuals interact in relation to the phenomenon under study
- Theory asserts a plausible relation between concepts and sets of concepts
- Theory is derived from data acquired through fieldwork interviews, observations, and documents
- Data analysis is systematic and begins as soon as data becomes available
- Data analysis proceeds through identifying categories and connecting them
- Further data collection (or sampling) is based on emerging concepts
- These concepts are developed through constant comparison with additional data

- Data collection can stop when new conceptualizations emerge
- Data analysis proceeds from "open" coding (identifying categories, properties, and dimensions) through axiel coding (examining conditions, strategies, and consequences) to selective coding around an emerging story line
- The resulting theory can be reported in a narrative framework or as a set of propositions

While we can amend or extend this list, it does convey some of the main tenets usually identified with grounded theory. As grounded theory becomes more popular and influential, we can expect summaries of this sort to become commonplace. Given the uncertainties that afflict social research and the knowledge it produces, there is something undeniably reassuring in the identification and repetition of key points setting out a methodological perspective. However, in this chapter, I suggest that a conventional summary of this kind provides only a very partial answer to the question: what is grounded theory? In fact, there are different and inconsistent answers, depending on where we look for one.

For example, we could look for an answer in the origins of grounded theory in the original text by Glaser and Strauss (1967), which launched grounded theory. But that was published more than three decades ago. So perhaps we should look at how grounded theory has evolved in the interim? Unfortunately, in doing so we find that there are now sharply divergent versions of grounded theory. By the 1990s, indeed, the original authors could no longer agree on what grounded theory was about. There can be few more scathing critiques than the one Glaser (1992) published of an updated version of grounded theory by Strauss and Corbin (1990). If the authors who inspired grounded theory have fallen out, it is not surprising to find some sharp differences of opinion among their disciples. Disagreements flare up, not only over what the methodology of grounded theory is in principle, but also over how to put into practice. Indeed, some critics dispute the claims of other researchers to have used grounded theory—not unlike, it may seem to an outsider, the way exponents of various cults bicker over the right interpretation of a religion.

Do the studies of researchers who claim to use grounded theory really have anything in common? When I suggested a book on grounding grounded theory to my publisher, I was asked which version of grounded theory I was going to discuss. Keen not to offend, my reply was (I hope) diplomatic: that there were probably as many versions of grounded theory as there were "grounded theorists." When even the major authors disagree over fundamentals, there seems little point in arguing over what is or should be regarded as the "correct" or "authorized" version of grounded theory. Rather than looking for an authorized version, we may find it more productive to explore the principles propounded and methodological issues raised in the name of grounded theory during the course of its evolution.

IN THE BEGINNING

The story that ended in apparent acrimony first began in 1967, when Barony Glaser and Anselm Strauss published a book titled *The Discovery of Grounded Theory*. This set out a manifesto for a different way of doing social research, with its own distinctive goals and methodology.

The new methodology was aimed at producing grounded theory, which was defined as "the discovery of theory from data" (Glaser & Strauss, 1967, p.1). The discovery of theory through the rigors of social research was contrasted with the discovery of theory from the comforts of the armchair. Being based on evidence, the former promised to generate theories that at least fit the immediate problem(s) being addressed as well as promising wider application. Glaser and Strauss suggested that theory must "fit the situation being researched, and work when put into use" (1967, p.3). There was more prospect of achieving this happy result, it was argued, if the theory in question was generated through research.

But how was this to be done? In order to generate theory through research, Glaser and Strauss proposed to use procedures for generating theory that differed from those required for testing it. We can sketch out these procedures in terms of the usual phases of the research process

- How to initiate the research
- How to select data
- How to collect data
- How to analyze data
- How to conclude the research

Let us look at each in turn.

HOW TO INITIATE THE RESEARCH

As theory was to be "discovered from data," the main point here was to avoid "preconceived" ideas. In grounded theory, Glaser and Strauss argued, "initial decisions are not based on a preconceived theoretical framework" (1967, p.45). The idea was to start instead with a general subject or problem conceived only in terms of a general disciplinary perspective. This would be a sufficient starting point for deciding what and where to study. Having identified a problem or topic in very general terms and selected a site where that problem could be studied, the researcher was then to allow the evidence they accumulated to dictate the "emerging" theoretical agenda. The first thing to do, in order to develop this agenda, was to identify "categories" which captured uniformities in the data, and then to identify their interesting properties and dimensions.

In this process of emergence, the researcher had to rely on their own "theoretical sensitivity" to generate relevant categories from the data. The researcher had to be able to think theoretically—to glean insights from the evidence, to conceptualize their data, and then to analyze relationships between concepts. Such "theoretical sensitivity" could be fostered by familiarity with "many different theories," but it also involved refusing to privilege any one theoretical perspective in advance of the ideas generated by the evidence itself. Thus

> An effective strategy is, at first, literally to ignore the literature of theory and fact on the area under study, in order to assure that the emergence of categories will not be contaminated by concepts more suited to different areas

This ignoring of the literature of theory and fact "at first" did not mean discounting the it altogether. Existing ideas could be applied—but only if warranted by the evidence:

> "A discovered, grounded theory, then, will tend to combine mostly concepts and hypotheses that have emerged from the data with some existing ones that are clearly useful." (Glaser & Strauss, 1967, p.46)

The important point, presumably, was not to be "doctrinaire"; the researcher had to explore evidence in its own terms rather than immediately fitting it into some preconceived framework.

We can summarize these points as follows (here and in the rest of the book, summaries are boxed off from the main text):

Initiating Research
• *Select an area of inquiry and a suitable site for study* • *Avoid theoretical preconceptions—ignoring the literature in that area (at first)* • *Rely on initial observations and "theoretical sensitivity" to develop categories and then relate them*

HOW TO SELECT DATA

Apart from the initial selection of a site for study, sampling in grounded theory could not be determined in advance of data collection. This could not be done in the initial stages of the research because sampling decisions must be theoretically informed and therefore await the emergence of a guiding theory:

> Beyond the decisions concerning initial collection of data, further collection cannot be planned in advance of the emerging theory (Glaser & Strauss, 1967, p.47).

Only once some theoretical ideas had emerged could the researcher determine what further data should be collected in order to explore and elaborate these ideas. Glaser and Strauss described this flexible and dialectical process of determining data collection in the light of the emerging analysis as "theoretical sampling"

> Theoretical sampling is the process of data collection for generating theory whereby the analyst jointly collects, codes and analyzes his data and decides what data to collect next and where to find them, in order to develop his theory as it emerges. This process of data collection is controlled by the emerging theory . . . (Glaser & Strauss, 1967, p.45).

In conventional methodology, sampling usually preceded analysis. Decisions would be made in advance—perhaps in the light of prevailing theory—about which population to sample or what agencies or events to study. In grounded theory the relationship was reversed: sampling decisions were to be based on the preceding analysis. Theoretical sampling had to be flexible, adapting to the emergent theory as it evolved. Glaser and Strauss contrasted this flexibility with the rigidity of conventional sampling strategies, which determined which agencies, sites, and individuals to include in the light of initial theory. This, they argued, prevented any adjustment of the data collection process to take account of new ideas thrown up by the research process.

There were other contrasts, too, with the conventional approach. Sampling governed by "theoretical" relevance was concerned with the contribution of new sources of data to conceptual development. This was where comparison came in. The crucial question concerned the selection of sites: which of these would provide appropriate comparable data to extend and deepen the emerging conceptualizations? In grounded theory the criteria for selection therefore revolved around comparison in terms of the concepts being investigated—rather than selection in terms of other factors which might delimit the populations or control the variables being studied. Comparability of sites (or populations) in terms of background variables such as location, function, or age was deemed irrelevant—or rather, could only become relevant in the light the emerging theory.

Comparison therefore proceeded in terms of the theoretical value of sites (for generating categories) rather than their representational value (as cases from which to generalize). For example, in studying awareness of dying, the authors selected for comparison those services that promised to illuminate key issues—such as those services where awareness about dying was high or low or where dying itself was rapid or slow. What mattered most in sampling was whether new data sources would offer interesting comparisons in terms of the processes being studied. What mattered least was the kind of service being studied; for example, whether the service to the dying was at home, hospice or hospital, emergency service, or intensive care. These would only be sampled—but any of them could be sampled—if they promised to illuminate (theoretically) the conditions under which nurse-patient interactions varied.

Selecting Data

- *Avoid sampling decisions before theory has emerged*
- *Focus on the contribution of new sites to theoretical development*
- *Select sites for comparison not cases for representation*
- *Select for conceptual variation across different sites*

HOW TO COLLECT DATA

Although by no means wedded exclusively to qualitative methods, grounded theory was presented largely in qualitative terms. Qualitative methods were judged most useful for exploring data and developing concepts, whereas quantitative methods tended to require preconceptualization of the data. Data for grounded theory were to be collected primarily through a combination of fieldwork methods, including observation, interviews, and documentary materials. Glaser and Strauss stressed the virtues of collecting data from a variety of sources, both to reveal variations and confirm conceptualizations.

At the outset of the inquiry, these varied methods of collecting data were liable to be unfocused. Data collection at this point could therefore be fairly comprehensive and achieved largely through unstructured methods. As a study proceeded, however, ideas would become more focused, and the methods of data collection could correspondingly become more structured. Interviews, for example, might resemble long conversations at the start of the study, but become highly selective and focused on particular topics (and therefore much shorter) by its close.

As in most other methods of qualitative research, data were to be analyzed as they were collected. A constant interplay (or dialogue) was envisaged between the data and the ideas that they generated.

Collecting Data

- *Use qualitative methods of data collection*
- *Look for a variety of data sources*
- *Start with unstructured methods of collecting data*
- *Develop a dialogue between data and analysis*
- *As theory emerges, use more focused methods of observation*

HOW TO ANALYZE DATA

Glaser and Strauss proposed a "constant comparative" method of generating and analyzing data. This method involves four "stages," that they characterized as (1) generating and (2) integrating categories and their properties, before (3) delimiting and then (4) writing the emerging theory.

The first stage involves identifying categories and their properties. The process of identifying categories is called "coding," and "incidents" initially could be coded under a multiplicity of categories. Categories (or codes) are to be generated by comparing one incident with another and then by comparing new incidents with the emergent categories.

A "category" is considered to "stand by itself" as a conceptual element of the theory; whereas a property is "a conceptual aspect or element of a category" (Glaser & Strauss, 1967, p.36). For example, "perceptions of social loss" was cited as a category that conveyed nurses" views of the degree of loss a death may involve for family and occupation. Another concept, "loss rationales" (the rationales that nurses use to justify their perceptions) was presented as a property of this category. Under the general heading of the "theoretical properties" of a category, Glaser and Strauss (1967, p.106) referred to types, continua, dimensions, conditions, consequences, and even its relation to other categories.

According to Glaser and Strauss, categories and their properties are concepts that must have two essential features. First, they have to be analytic—designating not entities per se but their characteristics. In other words, categories are not merely labels used to name different incidents but involved conceptualization of some key features. Categories are not considered by Glaser and Strauss as representations of the data but rather as "indicated" by it. Second, categories also have to be "sensitizing"—providing a "meaningful picture" that "helps the reader to see and hear vividly the people in the area under study" (Glaser & Strauss, 1967, pp.37–8).

Categories and their properties varied in their degree of abstraction. They were likely to emerge from the data at a lower level of abstraction, but as initial concepts were compared and contrasted, more abstract and integrating concepts might emerge:

> . . . different categories and their properties tend to become integrated through constant comparisons that force the analyst to make some related theoretical sense of each comparison (Glaser & Strauss, 1967, p.109).

This integrative process was supposed to reflect patterns of integration in the data itself (as generated through theoretical sampling); this provided the relevant data for revealing significant similarities and differences within and between categories. Although researchers could hypothesize about relations between categories, integration was certainly not to be "forced" on the data through the imposition of a pre-established model of how categories might relate.

Analyzing Data

- *Generate categories by coding observations*
- *Use categories that are analytic and sensitizing rather than representational*
- *Identify the properties of categories*
- *Develop integrative hypotheses about the relations between categories and their properties*

HOW TO CONCLUDE THE RESEARCH

In a sense, the research could be regarded as complete when written up—though Glaser and Strauss suggested that it was always possible for new points or perspectives to emerge even after an initial report has been produced. As the aim was to generate theory, the research could be concluded only when an adequate theory had emerged from the analysis. Knowing a theory was "adequate" also involved knowing when to stop collecting more data in terms of which to develop the analysis.

Glaser and Strauss characterized this point as one of "theoretical saturation" of the analysis. "Theoretical saturation" refers to the (non)emergence of new properties, categories, or relationships. Once the data no longer offer any new distinctions of conceptual import, categories could be described as "saturated" and no further evidence need be collected. Despite its metaphorical overtones of overwhelming excess (as in "saturation bombing"), this was not what Glaser and Strauss intended. If anything, they meant the exact opposite of a thorough and exhaustive trawl for and through the data:

> After an analyst has coded incidents for the same category a number of times, he learns to see quickly whether or not the next applicable incident points to a new aspect. If yes, then the incident is coded and compared. If no, the incident is not coded, since it only adds bulk to the coded data and nothing to the theory (Glaser & Strauss, 1967, p.111).

The data can be safely discounted once it has done its job of "indicating" ideas:

> Once a category or property is conceived, a change in the evidence that indicated it will not necessarily alter, clarify or destroy it . . . conceptual categories and properties have a life apart from the evidence that gave rise to them (Glaser & Strauss, 1967, p.36).

"Theoretical saturation" refers to concepts, not data, and identifies a point where no further conceptualization of the data is required.

While the limits of particular conceptualizations could be reached through theoretical saturation, the research as a whole could be brought to a conclusion through increasing selectivity. Through the process of integrating categories, a central theoretical framework could crystallize around a "core" category—otherwise described by Glaser and Strauss as the main "story line" of the study. A framework would "solidify" out of the analysis and delimit the research by differentiating between core and peripheral categories and identifying the scope and boundaries of the theory. This framework could in turn direct further data collection and analysis—but within a more circumscribed and focused agenda. The focus could shift to analysis of the major categories and their relations with efforts to identify underlying uniformities, clarify logical connections and "formulate the theory with a small set of higher level concepts" (Glaser & Strauss, 1967, p.110).

Writing up could then be done within the framework provided by the emergent theory. This provided the bare analytic bones of the research, with a decent padding provided by the memos written about the categories employed in the analysis. These could be brought together to provide a basis on which to amplify—and perhaps revise—the results of the analysis. The "coded data" could also provide a resource from which to draw in weighing evidence or offering illustrations of key arguments.

Concluding Research

- *Stop collecting data when they no longer produce significant conceptual variations ("theoretical saturation")*
- *Identify a "core" category or main "story line" for the study*
- *Integrate the analysis around this framework*
- *Use memos and coded data to amplify and modify the resulting analysis*
- *Stop when an adequate theory has emerged*

In later formulations, some of these analytic processes were discussed in more detail. Let us look now at some of the ways in which grounded theory has been elaborated or modified in subsequent accounts.

SUBSEQUENT ELABORATION

As grounded theory was presented as a "constant comparative" method, much of the subsequent elaboration of the methodology turned on a more detailed account of the process of "coding," which was seen by Glaser as the key process in constant comparison:

> The essential relationship between data and theory is a conceptual code. The code conceptualizes the underlying pattern of a set of empirical indicators within the data . . . the code is of central importance in the generating of theory (Glaser, 1978, p.55).

With reference to the "conceptual code," Glaser went on to distinguish between "substantive coding" and "theoretical coding." In this distinction, substantive codes were to "conceptualize the empirical substance of the area of research" (Glaser, 1978, p.55). If you like, these were "first-order" concepts, most closely related to data. Theoretical codes, on the other hand, were used to "conceptualize how the substantive codes may relate to each other as hypotheses to be integrated into the theory" (Glaser, 1978, p.55). These, then, were second order concepts, perhaps rather like the concept "furniture" is used to integrate various substantive categories such as "table" and "chair" into a wider classification schema. We can think of empirical examples of "tables" and "chairs," but "furniture" is a more abstract concept that obtains empirical instantiation only through the more substantive concepts to which it refers.

"Theoretical coding" established connections between categories. In order to render these explicit, the theorist had to know many "theoretical codes" which sensitized the analyst to the "implicit integrative possibilities in the data" (Glaser, 1978, p.73). Glaser listed 18 (overlapping) coding families that encompassed these possibilities, providing varied frames of reference in terms of which categories could be related. However, Glaser stressed that in selecting among these possibilities, analysts must always take their cues from the data.

"Coding" was further divided by Glaser into "open coding" and "selective coding." "Open coding" involves "coding the data in every way possible. . . for as many categories that might fit" (Glaser, 1978, p.56). "selective coding," by contrast, involves delimiting "coding" to only those variables that relate closely to the "core" variable that forms the heart of the emerging theory. Glaser (1978, pp.57–61) put forward a number of "rules" to insure the "proper use and success" of "open coding." These "rules" emphasized the analytic character of coding. Because coding is analytic rather than mechanical, you cannot delegate coding to a hired hand—you have to develop your own analy-

Rules Suggested by Glaser for Open Coding

- *Ask a set of questions of the data that must be kept in mind from the start:*
  What is this data a study of?
  What category does this incident indicate?
  What is actually happening in the data?
- *Analyze the data line by line*
- *Do your own coding*
- *Always interrupt coding to memo an idea*
- *Stay within the confines of the substantive area and field study*
- *Do not assume the analytic relevance of any "face sheet" variables (such as age, sex, etc.)*

sis. Because coding is analytic, you have to capture its conceptual significance by using memos to annotate the allocation of codes. Because coding is analytic, you have to concentrate on the nuances and subtleties of the data in detail by giving it the closest possible scrutiny.

Whereas Glaser set out a range of possibilities for integrating categories through "theoretical coding," Strauss and Corbin in their formulation (or re-formulation) of grounded theory focused on only one. This they described as "axiel coding," which is defined as:

> A set of procedures whereby data are put back together in new ways after open coding, by making connections between categories. This is done by utilizing a coding paradigm involving conditions, context, action/interactional strategies and consequences (Strauss & Corbin, 1990, p.96).

Like Glaser, Strauss and Corbin stressed the importance of "integrative" coding. But whereas Glaser recognized a range of potentially integrative concepts organized into a variety of different families, Strauss and Corbin selected one family (conditions, context, action/interactional strategies, and consequences) as a "coding paradigm." Strauss and Corbin claimed that the use of this coding paradigm allowed data to be related systematically in complex ways. Without it, researchers would be unable to achieve the desired combination of "density and precision" in developing their analysis (Strauss & Corbin, 1990, p.99).

Strauss and Corbin also proposed a "conditional matrix," which was defined as a "complex web of interrelated conditions, action/interaction, and consequences" pertaining to a given phenomenon (Strauss & Corbin, 1990, p.161). This matrix sets out a range of conditions or consequences from those immediately influencing the interaction, through to those at national and international levels that could (if relevant) be incorporated in the analysis.

The Elaboration of Grounded Theory

- *Distinction between substantive codes and theoretical codes (Glaser)*
- *Distinction between open coding and selective coding (Glaser)*
- *Use of coding paradigm in axiel coding (Strauss & Corbin)*
- *Use of conditional matrix to analyze conditions and consequences (Strauss & Corbin)*

FROM MANIFESTO TO MARCHING ORDERS

So far, all this may seem fairly sensible and straightforward (if rather general) advice, which, despite the later elaborations, might be reasonably be summarized as follows:

1. Find an interesting area of inquiry.
2. Select a relevant site.
3. Collect some data.
4. Generate some ideas.
5. Explore them through further comparisons.
6. Connect the emerging ideas.
7. Integrate them around a selected theme.

What could be more straightforward? But if matters were so simple, why the subsequent disputes over grounded theory? Let us look, therefore, at some of the problems that emerged as grounded theory evolved from an initial manifesto to a set of marching orders designed to elucidate its main procedures.

THE EVOLUTION OF GROUNDED THEORY

The initial text setting out grounded theory was a polemical work. It attacked the predominant modes of theorizing then current in sociology and set up grounded theory as an alternative approach. The polemic was directed against speculative or deductive forms of theorizing, in which theories were first dreamed up (preferably while resting comfortably in an armchair) and then subsequently "tested" against evidence through research.

The deductive mode cast the role of research primarily as that of "verifying" rather than generating theory, and the canons of verification were seen (at the time) as best met through the analysis of quantitative data produced through systematic and structured forms of research. Against this, Glaser and Strauss counterposed the need to relate theory more closely to evidence in the first place. This could be done through a more flexible research process in which the constant interplay of data collection and analysis provided a sound base for generating theory. One of their aims was to set out a logical framework for the conduct of this kind of approach. A related ambition was thereby to provide a "strong rationale" for the adoption and use of qualitative methods of inquiry (Glaser & Corbin, 1994, p.275). As we shall see later, each of these ambitions sowed the seeds of subsequent confusion.

Grounded theory at first evolved mainly in the area of health and especially nursing, where Glaser and Strauss themselves had initially applied it. The methodology was undoubtedly popularized partly through the impressive empirical studies published under its rubric, even though these seldom more than hinted at the methodological processes through which the results were produced. Face-to-face teaching and supervision had a more direct impact on the methodological diffusion of grounded theory, with a whole generation of "grounded theorists" emerging under the tutelage of the masters, particularly in nursing (Benoliel, 1996, pp.408–410). Using a medline search of articles keyed by "nursing" and grounded theory, Benoliel reported a sharp increase in articles related to grounded theory research in recent years (Table 1.1).

Of the 146 studies published in 1990–1994, 45% were published outside the United States compared with 15% in the preceding period, suggesting growing international interest in a grounded theory approach. I found the same pattern in a Bids search of social science journals in May, 1997. This showed 143 citations for the earlier period (1981–1994) growing to 225 for the years of 1995–1997. The adoption of grounded theory has gradually spread beyond its initial concentration in nursing, and, especially encouraged by its association with software for qualitative analysis, it is now making inroads into other practical fields (such as education) and other disciplines (such as psychology). Obviously this burgeoning of interest has not been based entirely on the apprenticeship/mentor model of knowledge diffusion. Others have taken up the approach on the basis of the initial book (Glaser & Strauss, 1967) and/or subsequent texts (Glaser, 1978; Strauss, 1987; Strauss & Corbin, 1990) elaborating the methodology. While the methodology has attracted wider interest, fears have been generated in the process that diffusion has led to dilution of its basic canons of inquiry. Popularity, of course, brings its own problems, evident in the disputes that have developed over the methodology and its application.

Ironically, Strauss and Corbin have themselves been castigated for "deviating from the original method" (Wilson & Hutchinson, 1996, p.123) in a paper confidently entitled *Methodological Mistakes in Grounded Theory.* Strauss is taken to task for not having understood the basic tenets of his own approach (Corbin, as a graduate student and subsequent colleague, could be excused). More tellingly, perhaps, a similar accusation against Strauss has also been made (and reiterated repeatedly) by Glaser. In response to the growing popularity of grounded theory, Strauss (now collaborating with Corbin) produced a basic introduction to grounded theory (Strauss & Corbin, 1990) intended to ensure that novice researchers understood just what was required of the methodology. Perhaps inevitably, this "codified" version has since become the standard introduction to grounded theory in

TABLE 1.1 Number of Articles Including Key Words "Nursing" and "Grounded Theory" in Medline Search

Period	Number of articles
1980–1984	5
1985–1989	33
1990–1994	146

Source: Benoliel (1996, p.412).

place of the original text. However, Glaser (1992) repudiated this "version," suggesting that it bore little relation to the original grounded theory. Indeed, Glaser suggested that in the revised form proposed by Strauss and Corbin it had evolved into a quite different methodology. Thus diffusion has brought dissension, and dissension (rightly or wrongly) has brought disrepute: "corruptions of the method in recent years place its credibility at risk" (Wilson & Hutchinson, 1996, p.124).

CURRENT ISSUES

Let us consider what issues these disagreements raise for any understanding of grounded theory.

The "deviation" from grounded theory denounced most emphatically by Wilson and Hutchinson concerns the introduction by Strauss and Corbin of the new coding paradigm involving conditions, context, action/interactional strategies, and consequences. The coding paradigm has also been criticised by Glaser, who likewise sees it as an (unacceptable) imposition of "pet theories" on data rather than letting theory emerge through the analysis. As this paradigm seems to impose a conceptual framework in advance of data analysis, it does not seem to sit easily with the inductive emphasis in grounded theory. This suggests that one issue in interpreting grounded theory lies in the scope it allows for adopting preconceived conceptual frameworks as an aid to analysis.

Wilson and Hutchinson (1996, p.123) also criticize the development of "rigid rules" for judging the value of grounded theory. Rather disparingingly, they attribute this development to "nurse scientists" with "no first-hand contact" with Glaser and Strauss or their students. In their ignorance, these "translators" (not apprentices) insist on methodological rules such as minimum sample sizes or diagrammatic presentations of theory. In insisting on such rules, they are condemned by Wilson and Hutchinson for committing a number of methodological sins. First, they break with "the spirit of creativity inherent in the original grounded theory method"—though Wilson and Hutchinson do not reject such rules entirely, for they go on to admit that "certain flexible methodological guidelines" are indispensable—and, indeed, they themselves go on to criticize some studies for failing to adopt particular guidelines.

A further criticism concerns "premature closure," by which Wilson and Hutchinson mean the failure to analyse data fully and especially to develop the more abstract "conceptual and theoretical codes" that they regard as the building blocks of theory. Some researchers fail to transcend an initial "in vivo" coding and so fail to move beyond the face value of their data. Another criticism concerns the importation of preconceptions (such as an established

A typical example of "grounded" theory" in nursing research is a study by Barclay et al (1997) using focus group interviews to collect data and grounded theory to analyze it. Note in the abstract below the reference to grounded theory only as an analytic technique rather than as a means of generating and interacting with data.

Barclay L., Everitt L., Rogan F., Schmied V., & Wyllie A. (1997) Becoming a mother—An analysis of women's experience of early motherhood. *Journal of Advanced Nursing, 25(4), 719–728.*

> This paper presents the results of a qualitative study conducted by midwife researchers into women's experience of new motherhood. Data were collected using focus groups involving 55 first-time mothers and analysed using grounded theory method. The analysis produced six categories: "realizing," "unready," "drained," "aloneness," "loss" and "working it out." The core category, "becoming a mother," integrates all other categories and encapsulates the process of change experienced by women. Also explained are factors mediating the often distressing experience of becoming a mother. The analysis provides a conceptualization of early motherhood enabling the development of strategies for midwives, nurses and others helping women negotiate this challenge.

conceptualization of "stages of dying") and hence a failure to develop "an original and grounded interpretation" (Wilson & Hutchinson, 1996, p.124). While we may be reluctant to conceded claims that studies making such "errors" are not using grounded theory, these criticisms certainly raise some doubts about the extent to which grounded theory does imply a specific, and agreed, set of methodological procedures.

A second, more abstract, strand of criticism advanced by Wilson and Hutchinson is that the translated "rules" for grounded theory also break with "the philosophy of pragmatism undergirding the method." Instead, it is suggested, "canons of quantitative, positivist method" are being adapted through grounded theory to qualitative methods of analysis:

> The outcome is a study report replete with conventional positivistic terminology, including random sampling, reliability and validity statistics, independent and dependent variables, and the like (Wilson & Hutchinson, 1996, p.124).

Other critics have even directed much the same criticism at the originators of grounded theory:

> Glaser & Strauss, in an effort to make themselves clear to quantitative methodologists, used the language of positivism: variables, hypotheses, properties, theoretical sampling, theoretical ordering and so on. It is often this discourse that causes frustration for the qualitative researcher (Keddy et al., 1996, p.450).

Another example of grounded theory in nursing research exemplifies its use in theorizing (Dildy, 1996). Note in the abstract below the reference to grounded theory only as a tool in data analysis without reference to its wider research role in generating as well as analyzing data.

Dildy, S.P. (1996) Suffering in people with rheumatoid arthritis. *Applied Nursing Research, 9*(4), 177–183.

> Knowledge of the nature, meaning, and impact of suffering from the perspective of people with rheumatoid arthritis is needed to determine what nursing interventions are most helpful in reducing suffering. Grounded theory was used to identify the nature of suffering in 14 people with rheumatoid arthritis. Suffering was found to be a process directed toward regaining normalcy and consisted of three phases: disintegration of self; the shattered self; and reconstruction of self. Experiencing suffering resulted in struggling, loss of dreams, restructuring a future orientation, and withdrawing. Finding meaning through positive life changes was an outcome of suffering. The informants' differentiation between pain and suffering also was examined. The provision of comfort measures along with a caring and empathetic attitude were identified as helpful nursing interventions in reducing suffering.

Indeed, these critics complain that Glaser and Strauss in their original text used language that is "esoteric and, contrary to the principles of grounded theory, sounds static and linear" (Keddy *et al.*, 1996, p.450). Thus even the originators of grounded theory have been castigated for writing "contrary to the principles" of their own methodology. How the mighty have fallen!

At any rate, these criticisms alert us to a further question about grounded theory: whether, as a methodology, it is (or should be) "grounded" in a particular "philosophical" approach. Thus Charmaz suggests that "the major problems with the grounded theory method lie in glossing over its epistemological assumptions" (1990, p.1164). However, this is perhaps hardly surprising, since she argues that "the relation between subjectivist and objectivist realities and levels of explanation remains unspecified" in grounded theory. Charmaz herself offers a "social constructionist" version of grounded theory which, she suggests, has a more phenomenological (and consistent) cast than the original:

> Glaser and Strauss' earlier works have both phenomenological and positivistic emphases and therefore, sometimes may seem confusing and even inconsistent. They claim to be phenomenological, yet a strong positivistic thread runs through their work (Charmaz, 1990, p.1164).

To Charmaz, the phenomenological slant in grounded theory is evident in the first-hand observation of issues in the real world; the positivistic slant on the other hand is evident in the implication that the real world dictated the shape of the emerging theory. Charmaz presents this as a "lapse"—though perhaps forgivable:

> Glaser and Strauss lean toward assuming that the theoretical categories derive from the data and that the researcher remains passive. Here, they come close to positing an external reality, unaltered by the observer's presence. Whether they intended to do so or simply had a theoretical lapse in the midst of methodological claims-making is itself open to construction (Charmaz, 1990, p.1164).

If lapse it is, it has still to be rectified. Charmaz acknowledges that Strauss had embraced the idea of "an actively involved researcher who constructs categories and concepts" in his more recent explication of grounded theory. Nevertheless, Strauss, in her view, remains "within the empirical science tradition" and "displays the tension between being simultaneously subjectivist and scientific" (Charmaz, 1990, p.1164).

The epistemological theme is taken up by another critic, Merilyn Annells, in her review (1996) of the relation between grounded theory and "post-modern" methods of inquiry. Whereas Charmaz contrasts "phenomenological" and "positivist" perspectives, Annells discards this dichotomy in favor of a fourfold typology of inquiry paradigms—positivism, postpositivism, critical theory, and constructivism—which is borrowed from Guba and Lincoln. Annells has difficulty in placing grounded theory in terms of any particular paradigm and sees this as a weakness rather than a strength.

The paradigms are distinguished by their answers to the following questions:

Ontological:	What is the form and nature of reality? What can be known about reality?
Epistemological:	What is the nature of the relationship between the knower (the inquirer) and the would-be knower and what can be known?
Methodological:	How should the inquirer go about finding out whatever he or she believes can be known?

Annells suggests that the way we answer the ontological question then shapes our answer to the epistemological question and that the way we answer the epistemological question then shapes our answer to the methodological question. But the answers to these questions offered (or merely implied) by grounded theory are ambivalent; moreover, they have shifted over time. Thus Annells contrasts Glaser's "classic" claim of grounded theory that theory "really exists in the data," with Strauss and Corbin's later insistence that reality "cannot actually be known, but is always interpreted" (Annells, 1996, pp.385–386). She also contrasts Glaser's ideal of coming "closer to objectivity" with Strauss and Corbin's recognition of the researcher as "a crucially significant interactant" in the research process (Annells, 1996, p.387). Finally, she contrasts Glaser's view that grounded theory generated through qualitative research can be subjected to subsequent verification with Strauss and Corbin's claim that verification is possible "throughout the course of a research project" (Annells, 1996, p.388).

Thus, in Annell's view, grounded theory is open to conflicting interpretations over some basic questions about the nature of reality, how it is to be apprehended, and what this implies for methodology. This "constructivist" criticism raises issues about the ontological and epistemological assumptions that inform grounded theory and the implications these may have for choosing between different methodological options.

Some Disputes about Grounded Theory

- *Can grounded theory presume a particular conceptual framework (such as the coding paradigm)?*
- *Must grounded theory transcend "in vivo" codes?*
- *Does grounded theory require a specific set of methodological procedures?*
- *Can researchers using grounded theory claim to be objective?*
- *Is grounded theory grounded in an external reality?*
- *Can grounded theory be verified as it is discovered?*

GROUNDED THEORY IN PRACTICE

Let us return from the hallowed heights (or depressing depths) of ontological and epistemological inquiry to more pedestrian concerns. I noted earlier Benoliel's survey of the expansion of research using grounded theory. Reviewing the abstracts of articles for the period 1990–1994, she suggested that only 33 could be interpreted as grounded theory research, while others were described as using a grounded theory approach which "made use of interview data only" and "did not account for social structural influences on the experiences of respondents" (Benoliel, 1996, p.412). This criticism reflects her view that the goal of grounded theory is "to explain how social circumstances could account for the behaviors and interactions of the people being studied" (Benoliel, 1996, p.413). By contrast, she notes, much of the work done in nursing has focused on analyzing the "lived experience of the client" rather than "social processes or passages over time." Benoliel also complains that the focus is often on social psychological processes viewed through a "narrow lens" while contextual elements at different levels (organization, community, or society) are less well delineated. These criticisms raise questions about whether grounded theory involves a focus on a particular kind of research question (such as Benoliel's stress on explanatory circumstances); on how grounded theory manages the relationship between action and context (or structure); and how it analyzes changes over time.

A related criticism suggests that grounded theory is often confused with other qualitative methods, such as phenomenology. This "method slurring" (Baker *et al.*, 1992) arises, it is argued, from the common misrepresentation of the different methodologies, leading to inconsistencies between the methods adopted and the underlying methodological assumptions. In this critique, grounded theory is presented as a qualitative method, whereas its inventors presented it as a method that transcended the qualitative/quantitative divide, having potentially equal relevance to either type of data—so long as the orientation was toward theory generation. This raises the question of how grounded theory does relate to qualitative research: is it a particular variant of qualitative methods or does it claim a more generic status?

Further Questions about Grounded Theory

- Must grounded theory address a particular set of research questions?
- How does grounded theory *manage the relationship between action and context?*
- *How does grounded theory analyze change over time?*
- *Is grounded theory more than one variant of a qualitative methodology?*

Dimensional Analysis

Another critique of grounded theory stems from a source rather closer to home. Schatzman taught with Strauss at the University of California but used a rather different method from grounded theory, which has since been described as "dimensional analysis." Schatzman claims of grounded theory that "the operations involved in discovering theory remained largely mysterious and undisclosed" (reported by Kools *et al.*, 1996, p.313). This, according to Schatzman, leads to a wide variation in the ability of students to use the methodology. He believes that there were "problems inherent in the original method" and, in particular, that "the absence of an overarching structure" impedes analysis (Kools *et al.*, 1996, p.314). This structural gap Schatzman aims to fill by a form of "natural analysis" derived from "the interpretive actions that one naturally and commonly employs everyday" (Kools *et al.*, 1996, p.314). A key element in natural analysis is "dimensionality," which:

> refers to an individual's ability to address the complexity of a phenomenon by noting its attributes, context, processes and meaning . . . The key process in dimensional analysis is the construction or novel reconstruction of the multiple components of a complex social phenomenon (Kools *et al.*, 1996, pp.315–316).

The procedures suggested by dimensional analysis are argued to differ in some detail from those of grounded theory in either its original or revised forms. For example, the initial emphasis is not on apprehending the "basic social process" in a situation, as suggested in grounded theory, so much as on analyzing its multiple dimensions—whereby "the researcher unravels and uncovers what all [sic] is involved in the phenomenon" (Kools *et al.*, 1996, p.317). Also dimensional analysis casts the "explanatory matrix" of context, conditions, processes, and consequences as a cornerstone of the analytic process and not just one among many techniques—though this seems to echo the proposed use of the coding paradigm by Strauss and Corbin. The precise details of dimensional analysis are less important, though, in my view, than the general questions it raises about how grounded theory is generated and whether and how it copes with the complexity of social situations/processes.

Glaser's Critique

Finally, and fittingly, let us turn to Glaser's critique of the "revisionism" he attributes to Strauss (and Corbin), some elements of which we have already touched upon above. One point raised by Glaser (and referred to earlier) concerns the role of "verification" in grounded theory. Glaser criticises the idea that grounded theory involves verification:

> Grounded theory is not verification. . . verification studies draw from a different methodology. . . . The hypotheses need not be verified, validated or more reliable. . . . The two types of methodologies should be seen in sequential relation (Glaser, 1992, pp.29–30).

In short, "verification" has no place in grounded theory. Let later researchers, who want to follow up the theory, worry about verifying it: "verificational work on or with the theory is left to others" (Glaser, 1992, p.16). On the one hand, Glaser argues, the task of grounded theory is to generate hypotheses, not to test them. On the other, he claims that verification is irrelevant precisely because ideas are induced from the data: "grounded theory looks for what is, not what might be, and therefore needs no test" (Glaser, 1992, p.67). Compare this view with the following comments by Strauss and Corbin:

> Regardless of level of theory, there is built into this style of extensive interrelated data collection and theoretical analysis an explicit mandate to strive toward verification of its resulting hypotheses (statements of relationships between concepts). This is done throughout the course of a research project, rather than assuming that

> verification is possible only through follow-up quantitative research (Strauss & Corbin, 1994, p.274).

The contrast could hardly be more striking. However, it is notable that Glaser himself accepted verification as part of grounded theory in his earlier discussion of coding:

> Open coding carries with it verification . . . Verification . . . is part of the delayed action nature of grounded theory so the analyst should not be misled by initial quick results (Glaser, 1978, pp.60–1).

Given such ambiguities and contrasts, we are left to puzzle over what role if any verification has in grounded theory research.

The disagreement over verification reveals a still deeper divergence between the originators of grounded theory. Glaser presents grounded theory as an inductive approach, that involves "figuring out from the patterns in the data what concepts and hypotheses emerge" (Glaser, 1992, p.71). There is not much room here, if any, for deductive inquiry, that involves inference in the other direction—from general propositions to particular instances. Deduction is viewed as the engine which drives verification, as we deduce consequences from our ideas which can be tested against evidence. Noting that Strauss and Corbin suggest a combination of inductive and deductive analysis, Glaser rejects the latter mode of inquiry as inappropriate in grounded theory because it is redundant. For there is no need to check theory against the data unless it is not based on the data but already goes beyond it. Thus we are left to consider whether grounded theory conforms strictly to the canons of inductive inquiry (whatever these might be) or instead involves some combination of inductive and deductive analysis.

This issue relates to another disagreement, this time over the role of prior theory. Strauss and Corbin (1994, p.277) dismiss the emphasis on induction in the original formulation of grounded theory as merely a form of "purposeful rhetoric" that "overplayed the inductive aspects." Instead they acknowledge

> both the potential role of extant (grounded) theories and the unquestionable fact (and advantage) that trained researchers are theoretically sensitized. Researchers carry into their research the sensitizing possibilities of their training, reading, and research experience, as well as explicit theories that might be useful if played against systematically gathered data, in conjunction with theories emerging from analysis of these data (Strauss & Corbin 1994, p.277).

Although couched in terms of "theoretical sensitivity," this refers not only "sensitized" researchers but also "explicit theories" which can be "played against systematically gathered data." Glaser by contrast emphasizes "emergence" of theory as the main feature of a grounded theory approach. He complains that playing existing theory against the data risks "forcing" the data to fit the theory.

This complaint is voiced most forcibly against the coding paradigm and conditional matrix that Strauss and Corbin propose as methods of conceptualizing the conditions and consequences of interaction. Glaser's criticism of the coding paradigm was noted above. The conditional matrix includes conditions at various levels of proximity to the interaction—ranging from close (e.g. the immediate setting) to the most distant (e.g. global conditions). Strauss and Corbin argue that conditions at all these levels are relevant to analysis, and the researcher "needs to fill in the specific conditional features for each level" (1990, p.161). Glaser, in contrast, regards this as merely the imposition of a set of "pet codes" on to the data. It involves asking questions, indeed, too many questions, of the data—questions that may not be at all relevant to emerging theory. It is far from clear, then, just what is supposed to be excluded by grounded theory in the way of "preconceived ideas."

These broad differences in interpretation also involve differences over procedure. For example, Glaser takes Strauss (and Corbin) to task for introducing sampling methods more appropriate to verificational studies. He argues that they shift the focus of grounded theory from the analyses of social processes to substantive generalizations about "units" of analysis. He also argues that the methods of data analysis—and in particular the replication of analysis through extensive coding—are more appropriate in verificational studies than in grounded theory. In conclusion, Glaser suggests that Strauss (and Corbin) have invented a new method, which he calls "full conceptual description." It is clear that he is not much enamoured of this alternative. He acknowledges it as "another method in its own right" but "it is surely not grounded theory" (Glaser 1992, p.62). This criticism raises further questions about the role in grounded theory of sampling techniques, methods of data analysis, and the scope and focus of generalizations.

Issues Raised by Glaser's Critique of Revisionism

- *What place has verification in grounded theory?*
- *Does deductive reasoning have a role in grounded theory?*
- *Can preconceived theories be "played" against data?*
- *What is the role of the coding paradigm and the conditional matrix in grounded theory?*
- *Can sampling in grounded theory be based on (representational) units of analysis rather than variation in a basic social process?*
- *Is extensive coding required in grounded theory?*

Whereas Glaser reasserts the authority of the original formulation of grounded theory against revisionism, Strauss and Corbin (1994, p.276) prefer to present grounded theory as a methodology in the process of evolution. They suggest that the approach has been influenced by contemporary trends, such as varieties of postmodernism, though it retains its core elements unaltered. The central features which they present as "constitutive" of the methodology are:

> the grounding of theory upon data through data-theory interplay, the making of constant comparisons, the asking of theoretically oriented questions, theoretical coding, and the development of theory (Strauss & Corbin, 1994, p.283).

However, as we have seen, this still begs a number of questions about how even these central features of grounded theory are to be interpreted.

SUMMARY

Some of the problems that plague social research have already become manifest in my brief account of the evolution of grounded theory. Chief among these is the intractable problem of "meaning" and its implications for how we conduct research. As we have seen, even such an apparently well grounded methodology as grounded theory has no uniform and self-evident interpretation. In this respect, meaning is like a bubble floating before our eyes: we see it plainly enough, but as soon as we try to grasp it, the bubble bursts and meaning evaporates. We have considered the original texts, subsequent reformulations, practical applications, and alternate approaches—and we are still no nearer to a clear interpretation which sets out what grounded theory means.

There is an irony—perhaps a paradox—here: that a methodology that is based on "interpretation" should itself prove so hard to interpret. Of course, there is no single, correct "interpretation," but a plurality of "interpretations." Do we accept the inspirational gospel according to Glaser? Or the safe schemas of Strauss and Corbin? Or perhaps Schatzman's doctrine of dimensionality? The religious metaphor may seem appropriate, given the way the initial "prophecy" has dissipated into a legacy of schism and scholasticism. It is in some respects a sad tale, if a familiar one, and it may have a silver lining. For it is surely our inclination to question "received authority" which breeds the divisions and disputes that mark (or mar) methodological evolution.

Even confining ourselves to a review of developments within grounded theory, we are left with a large number of questions to explore.

- How much scope does grounded theory allow for adopting preconceived conceptual frameworks as an aid to analysis?
- Does grounded theory imply agreement about a specific set of methodological procedures (and how fully specified should these be)?
- Does grounded theory need to be grounded in a particular epistemological and ontological approach?
- How does grounded theory relate to qualitative research methods?

- How does grounded theory conceptualize the relationship between action and context (or structure)?
- How does grounded theory analyze change over time?
- Does verification have any role in grounded theory?
- Does grounded theory allow deductive as well as inductive inquiry?
- Does grounded theory provide clear guidelines for sampling, data analysis and generalization?

Despite its immediate appeal—in the inspiration of the initial "manifesto," or the specificity of the subsequent "marching orders"—it turns out that grounded theory, and any claim to make use of it, raises more questions than it answers. This may be of considerable merit, since it is the very ambition to generate theory that is grounded in (mainly qualitative) data that forces us to confront some difficult issues. The very attraction of grounded theory may lie in the way it obliges us—because of its commitment to theory—to face up to some fairly basic issues about the nature of social research. If we accept the elementary (but awkward) principle that to do research requires reflection on what we are doing and how we do it, at the very least we should try to confront and clarify these issues.

CHAPTER 2

A Mixed Marriage

Though some contemporary researchers are seeking the solace of a "sacred" social science (Lincoln & Denzin, 1995, p.583), perhaps it is premature to abandon altogether a rational approach to methodology. Those taking the sacred or religious road tend to fall back on faith, while the rational path at least prompts us to press on with inquiry. Let us therefore press on with this inquiry into grounded theory by considering whether the later dissensions over grounded theory can be traced back to its earlier roots in the marriage of different analytic traditions. I suggest in this chapter that the questions raised earlier are not trivial issues, but rooted in unresolved tensions arising from the origins of grounded theory in rival traditions. These traditions were represented on the one hand, by Glaser, who brought to grounded theory the rigor associated with the quantitative survey methods of sociological research at Columbia University; and on the other hand, by Strauss, whose background lay in "symbolic interactionist" tradition of qualitative research as taught and practiced at the University of Chicago (Robtrecht, 1995, p.170). In the marriage of these two traditions, it was intended to harness the logic and rigor of quantitative methods to the rich interpretive insights of the symbolic interactionist tradition.

Although grounded theory brought together two very distinct traditions, subsequent interpreters have been more inclined to find its roots in "symbolic interactionism," while neglecting (or even bemoaning) the influence of the quantitative approach on the methodology. When critics argue that grounded theory needs to be understood in terms of its epistemological or ontological assumptions, it is the symbolic interactionist tradition of the Chicago school that they generally have in mind.

The Chicago school undertook field studies through which the researcher would "observe, record and analyze data obtained in a natural setting" (Robtrecht, 1995, p.170). The focus on behavior in natural settings reflected assumptions about the nature of social interaction:

> Human beings act towards things based on the meanings that the things have for them; the meanings of such things is (sic) derived from the social interaction that the individual has with his fellows; and meanings are handled in, and modified through an interpretive process and by the person dealing with the things they encounter (Bulmer, 1969, p.2 cited in Robtrecht, 1995, p.170).

The focus, then, is on both social interaction and its interpretation. Interaction is only possible through an interpretive process by which meanings are acquired or modified; interpretation in turn is acquired or modified through interaction. Therefore human behavior cannot be understood apart from the meanings that inform interaction. As these meanings emerge from interaction and are subject to continual revision, inquiry must study these meanings (and processes) as they evolve rather than treat them as "fixed" in (or, rather, out of) time.

Following an earlier discussion by Denzin, Silverman (1993, p.48) has summarized the methodological principles of "interactionism" as follows:

Principle	*Implication*
Relating symbols and interaction	Showing how meanings arise in the context of behavior
Taking the actor's point of view	Learning everyday conceptions of reality; interpreting them through sociological perspective
Studying the "situated" character of interaction	Gathering data in naturally occurring situations
Studying process as well as stability	Examining how symbols and behavior change over time and setting
Generalizing from description to theories	Attempting to establish universal interactive propositions

The interactionist approach itself developed in part as a critique of the assumptions of quantitative methodology, at least as practiced in the survey methods prevalent in mid-century American sociology. This was typically focused on the analysis of relationships among variables—such as those measured through attitude scales. It was argued that such variables tended to abstract from the complex interactions of everyday life—thereby reducing apparently complex social processes to simple (but simplified) relationships between abstract factors. Variables were often defined, it was argued, in terms that facilitated easy measurement rather than capturing the rich complications of human behavior. Moreover, "variables" required interpretations that were consistent and stable, whereas meanings tended to vary within individuals and among people as well as over time (Hammerlsey, 1989, pp.115–120).

How then can grounded theory marry the logic of variable analysis with naturalistic inquiry? If meanings are indeed shifting and volatile, how can they provide a solid basis for an analysis of variables and their interrelations? In their initial formulation of grounded theory, Glaser and Strauss do not address this question directly. They locate inquiry in naturalistic settings, focusing on interaction and its interpretation; and they construe analysis in terms of the identification of categories (variables?) and their relationships. They do not consider explicitly any possible inconsistency between the naturalistic inquiry and variable analysis.

However, there are some ways in which Glaser and Strauss do try to reconcile the potential tensions between their naturalistic form of inquiry and their variable mode of analysis. One approach they adopt is to treat concepts (or "categories") as though they can fulfil the functions of naturalistic inquiry and variable analysis. Categories are drafted in to serve as a bridge between the rich interpretive language of naturalistic inquiry and the formal concepts of quantitative analysis. A second approach Glaser and Strauss adopt is to suggest that theory can be grounded as it is generated. Thus theoretical concepts which are initially fluid can become fixed through the process of inquiry. A third approach is to present theory simultaneously as, on the one hand dense and complex, and on the other, simple and parsimonious. Thereby grounded theory can marry the requirements of naturalistic inquiry for rich description with the demands of variable analysis for reductive economy in the use of concepts.

In this chapter, I explore how well these various strategies succeed in resolving the problems arising from the marriage of two rather different and apparently competing traditions.

DUAL PURPOSE CATEGORIES

One strategy for reconciliation between variable analysis and naturalistic inquiry involves the adoption of the language of "categories" and "properties" in place of "variables" and "values." The latter do figure in the original account of grounded theory, but mostly in its (much neglected) application to quantitative data (Glaser & Strauss, 1967, pp.190–194). But if "categories" are to offer a means of mediating between transient interpretations on the one hand and stable conceptualizations on the other, they cannot simply be variables in another (qualitative) guise. They have to support both analytic roles. Therefore Glaser and Strauss (1967, pp.240–1) present categories as "sensitizing" concepts that relate meaningfully to the realities of interaction (as perceived by participants). At the same time, they are seen as "analytic" concepts designating "general properties" which form part of a theory. But unlike the abstractions of quantitative sociology, these are concepts that remain meaningful in the context of everyday interaction. Thus categories can bridge the gap, not only between but interpretive inquiry and variable analysis, but also between general theory on the one hand and its practical application on the other:

> The sociologist finds he has "a feeling for" the everyday realities of the situation, while the person in the situation finds he can master and manage the theory (Glaser & Strauss, 1967, p.241).

The concepts used by the analyst for analytic purposes are also "sensitive" to the social context in which they are used and applied. But how do categories attain (or maintain) this dual character? Can the analytic function in grounded theory, which requires stable concepts, really be reconciled with the sensitizing role, which suggests flexible interpretation? Let us consider in turn how Glaser and Strauss confront the problem of generating (stable) analytic concepts that also allow for sensitive (and flexible) interpretations.

Generating Stable Concepts

One striking feature of grounded theory is the essentially straightforward picture it presents of the analytic process. Glaser and Strauss seem to meet the first problem (of generating stable concepts) through the (implicit) elimination of diverse interpretations. They tend to underplay the diffusion, difficulties, and diversity of interpretations among those being studied. The action in grounded theory tends to be purposeful in character, with the actors guided by essentially "strategic" concerns. There is little sense of the deceits and self-deceptions, or conflicts and contradictions, that might make meaningful behavior so hard to interpret—even for the actors themselves. Though mediated by the researcher's own sensitivities, categories are "indicated" by the data in an essentially straightforward way. They do not have to compete for attention with other, rival ways of categorizing the data.

This interpretive stability is reflected in the injunction to avoid theoretical "pre-conceptions" that might prejudice interpretation one way or another. Glaser and Strauss seem to imply that categories can emerge from data almost without requiring any act of interpretation on the part of the researcher. The categories may be constructs, but these are constructs that directly reflect reality as experienced rather than preconceived.

Because categories are so plainly indicated by the data, they can then remain relatively undisturbed through the process of analysis. They may be extended, modified, refined or related—but rarely retracted. If interpretation is essentially straightforward, categories in grounded theory are rarely if ever mistaken with regard to existing data. They are merely modified in the light of new evidence. On this account, then, there is little need for continual interpretation and reinterpretation of the existing data. In grounded theory, the methodological thrust is toward the acquisition of further evidence, which may indeed shed new light on existing interpretations. Hence the emphasis on "constant comparison" with new sources of data acquired through "theoretical sampling"—rather than, for example, a methodological injunction to explore further the already available evidence germane to the (re-)interpretation of data already analyzed.

The stability imputed to emergent categories derives from the researcher's intimate knowledge of social realities as experienced by others, typically acquired through immersion "in the field":

> Why does the researcher trust what he knows? If there is only one sociologist involved, he himself knows what he knows about what he has studied and lived through. . .They are his perceptions, his hard-won personal experiences, and his own hard-won analyses (Glaser & Strauss, 1967, p.225).

And in an additional footnote:

> Researchers will readily agree that their own theoretical formulations represent credible interpretations of their data, which could, however, be interpreted differently by others; but it would be hard to shake their conviction that they have correctly understood much about the perspectives and meanings of the people whom they have studied (Glaser & Strauss, 1967, p.225).

Thus the researcher can rely on their "hard-won" experience to yield "correct" empirical understandings—even if their theoretical formulations may be open to alternative interpretations. This distinction between data and theory almost seems to imply a "pretheoretical" access to data such as the perceptions of others. This in effect would render the identification of stable meanings both reliable and unproblematic.

Nevertheless, Glaser and Strauss call upon practical action as a further basis for achieving "credible" interpretations of social interaction. These interpretations are tested by the requirements of having "in a profound sense, to live" in the social world being studied. The field workers have to "make out" in that social world to such an extent that they can:

> . . . quite literally write prescriptions so that other outsiders could get along in the observed sphere of life and action. . . If he has "made out" within the particular social world by following these prescriptions, then presumably they accurately represent the world's prominent features; they are workable guides to action and therefore their credibility can, on this account too, be accorded our confidence (Glaser & Strauss 1967, pp.226–7).

Credibility here becomes a by-product of practical reason—or, if you like, of common sense. Researchers as social actors have to "get by" in the social world being studied and, in doing so, can test the practical adequacy of their interpretations over time.

However, there seems to be a tension in grounded theory between this "immersion" in a social world—which is typical of the field worker who samples a single site—and the process of "theoretical sampling" which is required for comparative analysis. Theoretical sampling requires the inclusion of many sites, for these are needed to generate the necessary similarities and contrasts required by the emerging theory. There must be some doubt concerning the ability of the field worker to acquire "practical knowledge" of all these sites—at any rate within the compass of a single research project. To satisfy both requirements, it seems that both breadth and depth of knowledge are needed. Breadth is required to generate the needed comparisons, but depth is required to ensure the data are credible. It is not obvious how one can maximize both. The inclusion of many sites presumably reduces practical knowledge of any one of them. Inclusion of fewer sites presumably improves the practical knowledge of each, but at the expense of comparison. In practice, we may have to trade-off breadth against depth of knowledge.

Glaser and Strauss are nevertheless confident enough in the "credibility" of the data obtained through grounded theory that they dismiss any further attempts at improving credibility as a "compulsive scientism" that reflects the researcher's lack of trust in their ability "to know or reason" (1967, p.227). The researcher turning to additional methods of data collection (such as questionnaires) is admonished for "running away from his own ideas." Here again, the argument presented by Glaser and Strauss implies a straightforward and unproblematic relationship between the research experience and the ideas it generates. From this point of view, there is no need to trouble unduly over the "credibility"—or stability—of categories which are already grounded in experience. However, the research costs of acquiring that experience may be quite substantial.

Allowing Fexible Interpretations

If categories are stable, then how can they allow for flexible interpretation? If the social world is really in flux, then how can grounded theory grasp its fluidity while using concepts which are stable—and static? Glaser and Strauss cope with this problem partly through what might be called "theoretical flexibility." Theory is not stable but "evolves," and in the process it accommodates to and absorbs new information about the conditions and complexities of social interaction. Thus theory as a systematic set of interrelated concepts is not static, since the relationship between concepts is subject to continual adaptation and modification. New evidence rarely overthrows the original theory—instead it shows how to adapt or modify it to take this evidence into account. Thus theoretical progress is made through a smooth process of continual enrichment rather than marked by a staccato series of sporadic rejections and renewals.

Theory not only evolves; it is also interpreted. This interpretation varies according to the user, who must apply the theory to particular circumstances. In the process, the user can qualify the theory in various ways, for example, adjusting it "to fit the diverse conditions of different social structures" (Glaser & Strauss, 1967, p.232). Therefore grounded theory can be presented in general form, as its further and more detailed interpretation can be left to those seeking to apply it in particular conditions. Theory only has to be "good enough" to allow for such flexible interpretation:

> The presentation of grounded theory . . . is often sufficiently plausible to satisfy most readers. The theory can be applied and adjusted to many situations with sufficient exactitude to guide their thinking, understanding and research (Glaser & Strauss, 1967, p.235).

This flexibility in interpretation is contrasted with demands for more rigorous verification of grounded theory, especially where it has been generated through qualitative research. Rigorous verification requires time—and time is in short supply. Most researchers prefer to use this scarce resource to generate new theories (and careers!) rather than to replicate old ones. Glaser and Strauss recognized that social science has a short attention span. It is a matter of common knowledge that "once an interest in certain phenomena is saturated with substantive theory, attention switches to something else." This does not matter, though, if "the theory works well enough" so that users meantime "manage to profit quite well from the merely plausible work of discovery" (Glaser & Strauss, 1967, p.235).

Another reason theories may not (and need not) be subject to rigorous evaluation is that the pace of social change precludes it:

> . . . a great deal of sociological work, unlike research in physical science, never gets to the state of rigorous demonstration because the social structures being studied are undergoing continuous change. Older structures frequently take on new dimensions before rigorous research can be accomplished. The changing of social structures means that a prime sociological task is the exploration—and sometimes the discovery—of emerging structures. Undue emphasis on being "scientific" is simply not reasonable in light of our need for discovery and exploration amid very considerable structural changes (Glaser & Strauss, 1967, p.235).

Ultimately, then, the task is neither to evaluate theory nor even to amend it, but rather to generate new grounded theory to account for "emerging structures." Theoretical flexibility therefore involves not just modifications or reinterpretations of grounded theory in the light of new circumstances; it implies a fresh start to keep pace with the implications of social change.

From this perspective, it seems that any particular grounded theory is destined for a short shelf life. It can be consumed in any manner the customer sees fit and then dispensed with. This "short-term" view of theory may not square well with other aspirations often associated with theory, such as the accumulation of knowledge. Theoretical flexibility may accommodate to the transient nature of social reality, but at a price, perhaps, paid for by the limits this imposes on our theoretical ambitions.

Methodological Implications

Thus Glaser and Strauss reconcile the requirements of naturalistic inquiry and variable analysis by positing a dual role for categories as both analytic and sensitizing. In their analytic guise, categories provide the stability required to generate theory that transcends the immediate social context. In their sensitizing guise, categories offer the rich interpretations that are used in or can be applied to that context. In methodological terms, this is a demanding agenda. Sensitizing categories require immersion in the immediate social world so that they can be tested for practical adequacy through the requirements of everyday interaction. Analytic categories require extension beyond that immediate world to generate through comparison those patterns of similarity and difference upon which analytic purchase depends.

Glaser and Strauss reconcile these competing demands for depth and breadth in part by devolving the task of interpretation to the "user." They also minimize problems of interpretation at the conceptual level—that is, in the formulation of concepts—by emphasizing the stability of categories that have passed the test of practical adequacy. Categories that have once been "indicated" by the data are not likely to be discarded later. And they stress the importance of flexible interpretation, not so much at the conceptual level as at the level of theory—that is, in the relationships between concepts. We might compare this approach to that of a ball game in which the patterns of play may vary as the game unfolds, but the players themselves remain essentially the same. We may have to revise their properties as they show different skills and capabilities under changing conditions. Or we may have to play them in different positions. But they remain as key figures in a constantly evolving process—at least until another game starts requiring the generation of a new team of players.

This analogy prompts a question about the possibility of making substitutions during the course of play. We might like to change some of our concepts, and not just their relationships, as our research progresses. Pursuing the analogy, we might also like to start a new game with some familiar players rather than create an entirely new team. In other words, we might like to call upon some familiar concepts to generate grounded theory rather than always start from scratch.

The dual role of categories as analytic and sensitizing does not appear to entirely resolve the problems of marrying interpretive inquiry and variable analysis. The test of practical adequacy may seem a stringent one in terms of research costs, but even so it is not stringent enough. After all, it is not impossible that one can "get by" in a social world without having a valid appreciation (or interpretation) of its nature. Therefore the interpretive role of categories remains problematic. On the other hand, the flexibility acquired through insistence on generating new theory seems to preclude the kind of conceptual stability that must be attained if research is to "progress" and not continually to start from scratch.

Some Questions about the Dual Role of Categories

- *If categories are "indicated" by data does that imply that they are somehow pre-heoretical?*
- *Is a test of "practical adequacy" through intimate experience of the social world being studied compatible with a methodology of "constant comparison"?*
- *Should concepts be as subject to revision and reinterpretation as the connections between them?*
- *Should interpretations of evidence not always be treated as provisional?*
- *Does the generation of new theory require the disregard of prior theories?*

DISCOVERY

So far we have considered the use of sensitizing and analytic categories as a means of reconciling the tensions between "naturalistic" inquiry and "variable" analysis. Another way in which grounded theory handles these tensions—at least implicitly—is through the language of "discovery."

In the title of the original text, Glaser and Strauss proclaimed *The Discovery of Grounded Theory*. On the very first page, they define grounded theory as "the discovery of theory from data" (Glaser & Strauss, 1967, p.1). However, this leaves open whether it is theory that is being "grounded," or merely the process of discovering it. Glaser and Strauss seem to make two claims, one more and one less ambitious.

The less ambitious claim concerns grounding the process of discovery. It asserts only that the process through which theory is "discovered" from data is more "grounded" than that in which theory is "discovered" (or better, "invented") by the armchair theorist. Glaser and Strauss thought there were rather too many armchair theorists around in the 1960s, and they also thought the resulting theories were mostly unimpressive. Better theories could be "discovered" through social research than through abstract and speculative processes of logical deduction from ungrounded premises. The "discovery" of theory through research promised to open up much more fruitful lines of inquiry. In this claim, therefore, it is the process of discovering a theory—rather than the theory itself—that is "grounded." From this perspective, a theory might still need be grounded through various forms of subsequent "verification"—that is, by testing it against evidence—regardless of how it was first discovered. So a grounded theory from this standpoint would be a theory that has stood the test(s) of time.

The more ambitious claim asserts that not just the discovery process, but the resulting theory too is grounded through the way it is discovered. Thus the originators of grounded theory claimed that "the adequacy of a theory for sociology today cannot be divorced from the process by which it is generated" (Glaser & Strauss, 1967, p.5). This is a stronger claim because there seems little point in testing a theory which has already been "grounded" through the process of producing it. The argument is that a theory generated from data is grounded in a way that armchair speculation is not. As we have just seen, there are various reasons advanced as to why grounded theory may not or need not be rigorously tested against evidence. Moreover, Glaser and Strauss (following Kuhn's analysis of how science evolves) suggest that theories are rarely "falsified" by evidence—they tend to survive until a better theory comes along. So Glaser and Strauss argue (1967, p.4) that theories generated from data generally cannot be completely refuted or replaced: they are "destined to last." In this more ambitious claim, then, it is not just the process of "discovering" theory, but also the theory itself that is thereby "grounded."

These differing interpretations raise a question about the status of grounded theory with regard to discovery and verification. In grounded theory it seems that the two processes of discovery and verification are sometimes distinguished and sometimes conflated. At times, discovery is presented as a process that includes verification—or, at least, as much verification as a theory requires to be of practical value. At other times verification is presented along conventional lines as a follow-up to discovery. This approach presents research as an on-going process in which theory is first "discovered" and then tested. From this perspective, a theory can be grounded only if it withstands the test(s) of further inquiry.

The Logic of Discovery

In social research it is common to conceive of discovery and verification as separate activities, each with its own logic. Discovery usually implies a process of creative exploration and may be governed by quite different guidelines than those applied to verification, where we want to examine the validity of our ideas. Exploration favors freedom in the invention of ideas while examination requires rigor in testing them against evidence.

Of course, if our theoretical discoveries are grounded in the research process, our theories presumably stand more chance of later verification. Though this would seem to be an essential element in the argument for grounded theory, it is one that is surprisingly underdeveloped. In fact, Glaser and Strauss make very little direct comparison of their own with other methods of generating theory, such as the thought experiment. They do not explicitly consider, for example, whether other methods of theorizing (other, that is, than engaging in social research along the lines they suggest) might be more conducive to the creative process. The armchair theorist is dismissed, but without much consideration of the creative potential that we usually associate with free thinking—if not in the armchair, then at least in the bathtub or on the back of the envelope.

This bias against creativity is reflected in the language of discovery, which implies that theories are revealed rather than contrived. You cannot discover a theory in an armchair. But if theories are contrived rather than revealed, then it is not clear that you can discover a theory at all, even through social research. Despite all the stress Glaser and Strauss place on theoretical sensitivity, the language of discovery actually seems to downplay the creative element in theorizing.

This devaluation of creativity is often reinforced by the way Glaser and Strauss present the analytic process in passive mode. For example, they write of categories that "the diverse properties themselves start to become integrated" (1967, p.109) or again that "appropriate hypotheses will develop and quickly integrate with each other" (1967, p.171). This use of the passive voice seems to write the researcher entirely out of the picture. Concepts, properties, and their relations seem to "emerge" almost automatically from the data—though Glaser and Strauss do argue that the researcher does need some "theoretical sensitivity" in order to recognize and register this emergence.

At times, though, the researcher is cast in a more active role:

> When he begins to hypothesize with the explicit purpose of generating theory, the researcher is no longer a passive receiver of impressions but is drawn naturally into actively generating and verifying his hypotheses through the comparison of groups (Glaser & Strauss, 1967, p.39).

It is notable that the researcher becomes active not so much in conceptualizing data as in making connections between concepts. This activity also tends to focus on the further collection and comparison of data to extend the process of connecting concepts. Even here, apart from selecting groups and methods of data collection, the researcher's role in this process of comparison can be presented as essentially passive, merely recognizing and recording the points of similarity and difference that emerge from theoretical sampling. Even the emergence of a core theory is sometimes presented in largely passive terms:

> . . . as categories and properties emerge, develop in abstraction, and become related, their accumulating interrelations form an integrated central theoretical framework—the core of the emerging theory. The core becomes a theoretical guide to the further collection and analysis of data. Field workers have remarked upon the rapid crystallization of that framework, as well as the rapid emergence of categories (Glaser & Strauss, 1967, p.40).

This seems to imply that theories are shaped by the emergent "facts" in a fairly unambiguous fashion.

Glaser and Strauss do recognize that the "facts" themselves may vary:

> As everyone knows, different people in different positions may offer as "the facts" very different information about the same subject, and they vary that information considerably when talking to different people. Furthermore, the information itself may be continually changing as the group changes, and different documents on the same subject can be quite contradictory. . . (Glaser & Strauss, 167, p.67).

Nevertheless, they argue that the researcher, by comparing different views, can reach "a proportioned view of the evidence" as "biases of particular people and methods tend to reconcile themselves as the analyst discovers the underlying causes of variation" (Glaser & Strauss, 1967, p.68). The method of "constant comparison" can therefore provide a solid foundation for "confidence in the data upon which he is basing his theory" (Glaser & Strauss, 1967, p.68). Hence, perhaps, the transposition of "discovering facts" to discovering rather than constructing theory.

> Discover: *Expose to view, reveal, make known, exhibit, manifest; find out facts; become aware*
> Construct: *Fit together, frame, build, combine, draw, delineate*
> Fact: *Thing certainly known to have occurred or be true, datum of experience; what is true or existent; reality*

The language of "facts" reinforces the idea of discovery as revelation rather than invention. We usually speak of discovering facts rather than theory—rather as we may speak of Columbus "discovering" America. Nowadays, we might see the claim that Columbus discovered America as contentious, in terms of both its ethnocentric lack of vision—since native American Indians were already there—and its historical inaccuracy—since Columbus struck land in the West Indies, and other explorers probably reached the North American continent before him. So the facts are certainly subject to interpretation, depending on what claims are being made. Nonetheless, in the sense that something that was unknown to the European medieval world then became known—that is, within the context of European society at the time—the claim that Columbus discovered America seems indisputable. (I am not claiming it as "certainly true"—other than in the sense that some "facts" are so well established that they are not worth disputing).

Such "facts" are disclosed, it is presumed, through more or less incontrovertible evidence produced by experience or action—they are not "invented" (which is not to deny that they may first have to be imagined). And the language of "discovery" implies a high degree of certitude, as though these "facts" are established beyond reasonable contention. They cannot be "undiscovered." To claim that grounded theory is "discovered" may therefore impute a (false) finality to the theories we produce. It is no coincidence, perhaps, that Glaser and Strauss claim that "the truly emergent integrating framework . . .[is] hardly subject to being redesigned" (1967, p.41).

There is a strong tendency, therefore, to dispense with verification altogether. As we have seen, Glaser and Strauss claimed that further testing of an already plausible theory might be rendered irrelevant either by the predilections of social researchers (why bother!) or by the pace of social change. This does not preclude verification in principle—however unlikely it might be in practice. On the other hand, the claim that theory is already "grounded" through the process of discovery suggests that in some sense it has already been verified.

This implication is reinforced by the suggestion that concepts grounded in (or indicated by) data may be amplified or modified, but rarely discarded, in the light of new evidence. It is reinforced still further by the claim that the constant comparative method provides a means of "testing" hypotheses against evidence. At times, Glaser and Strauss quite explicitly embrace verification as part of the process of generating theory, though when they do so they accord it a subordinate status:

> . . . verifying as much as possible with as accurate evidence as possible is requisite while one discovers and generates his theory—but not to the point where verification becomes so paramount as to curb generation (Glaser & Strauss, 1967, p.28).

Thus discovery "subsumes and assumes verifications and accurate descriptions"—so long as these are subordinated to the task of generating theory. There need be no sharp separation of discovering and verifying theory:

> We would all agree that in social research generating theory goes hand in hand with verifying . . . Surely no conflict between verifying and generating theory is logically necessary during the course of any given research (Glaser & Strauss, 1967, p.2).

In practice, however, Glaser and Strauss do recognize logical conflicts between generating and verifying theory; and when they do so, they appear to give precedence to the former.

This ambivalence in grounded theory about the status of discovery and verification is evident in its methodological prescriptions. Let us look, for example, at the procedures of theoretical sampling and saturation.

Theoretical Sampling and Saturation

With theoretical sampling we are assured that rigorous sampling methods are not relevant to the selection of sites or subjects, since the aim of the exercise is to generate new points of comparison for developing theory. Since the aim is to generate theory, rather than to apply it to particular agencies or populations, we need not be unduly concerned with random sampling or similar procedures designed to provide a firm basis for generalization. Thus the logic of discovery justifies procedures that maximize the production of new ideas. But the same procedures do not provide a strong basis for generalizing these ideas to particular populations.

With saturation Glaser and Strauss propose a procedure for stopping data collection or analysis when no new ideas are being generated by the data. The saturation criterion is explicitly contrasted with procedures for testing analysis against all the data or weighing how far the evidence supports an emerging conceptualization. We are warned against the folly of finding further data against which to test categories that have already become saturated. We are reassured that the collection and analysis of data can become increasingly selective as the research becomes more focused. Such shortcuts are justified by counterposing the freedom of discovery against the rigors of verification. It is the generation of ideas that counts and not the full examination of supporting evidence. Indeed, even if a sample is biased or evidence inaccurate, Glaser and Strauss argue that it will have served its purpose if it stimulates conceptualization.

In practice, then, the logic of discovery is used to justify procedures that are incompatible with verification through the rigorous collection and analysis of evidence. This is not acknowledged, however, in the argument that "no conflict between verifying and generating theory is logically necessary during the course of any given research (Glaser & Strauss, 1967, p.2). Nor is it acknowledged in the assumption that by grounding the process of discovery in social research, we can also produce grounded theory—that by generating theory from data, verification can be accomplished in the same process. Of course, this requires some relaxation of standards of proof, but that is a requirement over which Glaser and Strauss themselves seem quite relaxed.

To recap: two quite different conceptions of grounded theory seem to be used. In the less ambitious, the logic of discovery is divorced from that of verification; while in the more ambitious, the logic of discovery and verification are married. Where they are divorced, grounded theory claims to offer no more than a better way of generating good ideas than the armchair (or the bathtub). Where they are married, grounded theory claims to offer a way not only of generating good ideas but also of verifying them. However, these two conceptions seem to be quite at odds, at least in their methodological implications. The emphasis on discovery through data seems to downplay the creative process in the generation of theory, while the presumption of verification seems to relax the requirements for rigorous testing of emergent theory against evidence.

Some Questions about the Discovery of Grounded Theory

- *Is it the process of discovery that is grounded, or theory as the product of that process?*
- *If theory is "discovered," does this imply that it is revealed rather than created?*
- *Is the generation of theory through empirical inquiry more creative than other forms of theorizng?*
- *Does the language of "discovery" tend to discount problems of interpreting the "facts"?*
- *Is it reasonable to relax the requirements for verification if theory is generated through empirical inquiry?*

THEORIZING

Another dilemma raised by the marriage of naturalistic inquiry with variable analysis concerns the kind of theory we are trying to generate. Are we trying to develop theories that reflect the complexities of social life and are therefore limited in scope, rich in detail, and bounded by context? Or are we trying to develop theories that simplify that complexity, and are therefore broad in scope, uncluttered by detail and unbounded by context? Naturalistic inquiry, with its emphasis on shifting meanings and motivations, favors the former. Variable analysis, interested in generalization and prediction, favors the latter. Glaser and Strauss seem to favor both.

Substantive and Formal Theory

One way they try to reconcile these divergent demands is through a distinction between two kinds of theory, which they call "substantive" and "formal" theory:

> By substantive theory we mean that developed for a substantive, or empirical area of sociological inquiry, such as patient care, race relations, professional education,

> delinquency, or research organizations. By formal theory, we mean that developed for a formal, or conceptual, area of sociological inquiry, such as stigma, deviant behavior, formal organization, socialization, status congruency, authority and power, reward systems or social mobility (Glaser & Strauss, 1967, p.32).

Both of these kinds (or levels) of theory can be generated through comparative analysis, though with differences in the strategy of inquiry required. Substantive theory requires a comparative analysis "between or among groups within the same substantive area." In contrast, formal theory requires comparative analysis "among different kinds of substantive cases which fall within the formal area" (Glaser & Strauss, 1967, p.33). This contrast between analysis within the same substantive area or across different substantive areas is illustrated by their work on dying as a "non-scheduled status passage." For example, "dying" is deemed a "substantive" area; and a comparison across hospital wards where patients die at different rates provides a basis for generating "substantive" theory. On the other hand, a theory of "status passages" is a "formal" theory, aspects of which can be generated by comparisons across substantive contexts, for example by comparing spies and building subcontractors with dying patients.

Glaser and Strauss argue that formal theories should not normally be applied directly to data, lest this involve forcing data to fit the theory. Instead, the research should allow substantive theories to emerge first, and on that basis determine which formal theories might be most relevant. This allows for "a progressive building up from facts, through substantive to grounded formal theory," thereby contributing to "the cumulative nature of knowledge and theory." Ultimately, the aim is "to arrive at more inclusive, parsimonious levels" through comparison of multiple theories at substantive and formal levels (Glaser & Strauss, 1967, p.35). Nevertheless, its close links with substantive theory mean that formal theory is always grounded in and not imposed on data.

Thus substantive theories can handle the details of empirical diversity while formal theories are more inclusive and parsimonious. The complexities and richness of social life are grasped at a substantive level:

> Using the constant comparative method makes probable the achievement of a complex theory that corresponds closely to the data, since the constant comparisons force the analyst to consider much diversity in the data (Glaser & Strauss, 1967, p.114).

On the other hand, even substantive theories must involve some level of abstraction:

> To make theoretical sense of so much diversity in his data, the analyst is forced to develop ideas on a level of generality higher in conceptual abstraction that the qualitative material being analyzed. He is forced to bring out underlying uniformities and diversities, and to use more abstract concepts to account for differences in the data (Glaser & Strauss, 1967, p.114).

Thus the difference between substantive and formal theories lies in their degree of conceptual abstraction. We can move from one to the other by focusing on "a higher level of generality" and incorporating material from other substantive areas "with the same formal theoretical import (Glaser & Strauss, 1967, p.115).

Unfortunately, once we acknowledge that the difference between them is a matter of degree, any dichotomous distinction between substantive and formal theories is hard to maintain. It may make more sense to talk instead in terms of degrees of generality or abstraction. Even Glaser and Strauss seem to have difficulty applying their own classification, as the following comment indicates:

> The substantive theory also could be generated by comparing dying as a status passage with other substantive cases within the formal area of status passage, where scheduled or not, such as studenthood or engagement for marriage. The comparison would illuminate theory about dying as a status passage (Glaser & Strauss, 1967, p.33).

This could just as easily be read as an example of "formal" theorizing about status passage through a comparison across substantive areas. There seems no "substantive" connection between engagement and dying other than through their conceptualization as status passages. Indeed, it seems hard to imagine a substantive theory that could be entirely devoid of conceptual content and yet in some sense claim to be a theory. Even dying has to be conceptualized as a social process—to be distinguished, for example, from sudden death—or, for that matter, from life itself, since dying is part of living. On the other hand, it may be equally hard to imagine a formal theory devoid of empirical reference. The concept of status passages may be a sociological abstraction, but it has empirical roots in the social processes and rituals that attend key life events.

In short, the distinction between formal and substantive theory might be better recast in terms of degrees of abstraction in which theory at any level has some combination of both substantive and formal elements. This still allows for a distinction in terms of theoretical emphasis, for theory may focus on either capturing the complexities of specific cases or on the generating (or condensing) generalities across a range of cases. However, it also implies a less clear-cut division between the theoretical tools to capture individual complexity and those we might use to make generalizations. As I argue later, we may use generalizations in analyzing the complexity of the individual case, while we also have to generate and apply generalizations within specific contexts. Glaser and Strauss do hint at this more dialectical conception of theory in their account of the dynamic interplay of substantive and formal theory, despite the rather rigid distinction they draw between the two.

Reduction and Focus

The tension between the demands of complexity and parsimony is also evident in the procedures Glaser and Strauss recommend for generating substantive theory. Considerable stress is laid upon the conceptual diversity of the categories and their properties and relations, which can be generated through constant comparison. We are told that "the truly emergent integrating framework . . . encompasses the fullest possible diversity of categories and properties" (Glaser & Strauss, 1967, p.41). Note that it is conceptual diversity that is the ambition here—not richness of empirical detail, so much as richness in conceptual variation. On the other hand, the production of maximum diversity may threaten the researcher with conceptual overload. This is because grounded theory seems to aim at both full contextual awareness of particular cases allied to full analysis of variation among a range of variables across a wide range of cases.

How does a grounded theory which maintains "a close correspondence of theory and data" achieve what Glaser and Strauss describe as "two major requirements of theory"—namely parsimony in formulation and generality in scope? Glaser and Strauss suggest two ways of "delimiting" theory. The first is through "reduction":

> By reduction we mean that the analyst may discover underlying uniformities in the original set of categories or their properties, and can then formulate the theory with a smaller set of higher level concepts. This delimits its terminology and text (Glaser & Strauss, 1967, p.110).

As an example, they cite the identification of a uniformity underlying the "loss rationales" of nurses in their focus on a patient's loss of social value (should they live); and then the generalization of that uniformity, from nurses dealing with dying patients to the social values of professionals distributing services to clients. The "reduction" is achieved first by through substituting "professionals in general" for "nurses," and secondly through substituting service distribution to clients in general for care of the dying in particular.

This reduction, however, seems to commit what is usually deemed a cardinal sin (in grounded theory) of going beyond the data. The evidence suggests only that some professionals (nurses) use loss rationales in some contexts (patients are dying). Even if we use constant comparison to examine the way other professional groups distribute resources to clients, the generalization to all forms of professional-client distribution still goes well beyond the available data. This may be acceptable from the point of view of generating theory, but it raises further doubts about how far that theory is grounded in the data.

Another way in which Glaser and Strauss try to obviate the danger of conceptual overload is through the identification of core variables around which to organize the analysis. This seems to be a simplifying device which allows the analyst to prioritize among variables and thereby to bring the analysis of variation within reasonable proportions:

> The second level for delimiting the theory is a reduction in the original list of categories for coding. As the theory grows, becomes reduced, and increasingly works better for ordering a mass of qualitative data, the analyst becomes committed to it. His commitment now allows him to cut down the original list of categories for collecting and coding, according to the present boundaries of his theory. In turn, his consideration, coding, and analyzing of incidents can become more select and focused. He can devote more time to the constant comparison of incidents clearly applicable to this smaller set of categories (Glaser & Strauss, 1967, p.111).

However, one of the puzzles posed by grounded theory concerns how this selection of core variables is to be accomplished. Once again, the language seems to eliminate the analyst from this process, other than through his or her commitment to what emerges from the analysis. There seems little conception here of choice among options, or of maintaining alternative hypotheses throughout the course of an analysis.

At whatever stage in the research process one variable becomes "privileged," the problem arises that data suggesting alternatives may be ignored. By focusing on a single core variable, the research agenda may become one dimensional rather than multi-dimensional. In effect, the analysis is "foreclosed." There seems to be a conflict here between a continually open-ended, multivariate analysis of relationships among all relevant variables and an increasingly selective focus on one main variable. This procedure promises to make the analysis more manageable, but perhaps it can no longer deliver "the fullest possible diversity" of categories and their properties.

Thus the problem of producing theory that is complex and parsimonious is not so much resolved as recast in a new guise. The distinction between substantive and formal theories allows complexity in the generation of substantive theory to be condensed into a more parsimonious formulation at a formal level. Theoretical reduction allows the elimination of superfluous specificity in the construction of generalizations. Focusing on a core category or variable allows the researcher to set some boundaries to the analysis. However, we may question whether any of these strategies really eliminates a trade-off between complexity and parsimony in the process of conceptualization.

Some Questions about Theorizing

- *Can a distinction between substantive and formal theory be sustained?*
- *Are complexity and parsimony characteristic of theories at all levels?*
- *Can theoretical reduction be achieved without loss of complexity?*
- *Can core categories be selected without reducing diversity?*

IN CONCLUSION

In light of this discussion, the confidence with which researchers claim to use a grounded theory methodology may seem surprising. On the other hand, the diversity of studies assuming the methodology may seem less so. Recalling the disputes noted in the previous chapter, it seems that some of the later difficulties and disagreements over grounded theory can be located in the ambiguities of the original text. For example, the later argument over the role of verification in grounded theory can be traced to ambiguities in the original presentation. Sometimes grounded theory is presented only as method of generating theory and sometimes it is presented as a way of generating and also verifying theory. The argument over the role of prior conceptualization in grounded theory can likewise be traced to the problems of reconciling a particularistic focus on generating substantive studies with the integrating claims of more formal theory. An emphasis on the "emergence" of theory is easier to maintain in the context of substantive than of formal theorizing.

Glaser and Strauss set out to mark a course between the contrasting extremes of impressionistic qualitative studies and rigorous quantitative analysis. This course attempted to marry the richness and complexity of the former with the scope and simplicity of the latter. However honorable the aim, the task emerges as perhaps more onerous than the authors realized. We are told that categories must be sensitizing and analytic, but it is less clear how they can be both. We are told that theory can be grounded as it is generated, but it is not clear how this can be accomplished. We are told that analysis should be both simple and complex, but it is unclear how this is to be done.

In this chapter, my aim has been to raise questions rather than to resolve them—if indeed they are capable of resolution. However, the discussion thus far already suggests some of issues raised by grounded theory which might need further consideration. One concerns the status of categories and what confidence can be placed on them in the processes of conceptualization. Another concerns the role of creativity in the generation of theory, which Glaser and Strauss suggest is fostered through the disciplines of empirical inquiry. Another issue concerns the extent to which grounded theory can claim to be validated through methods by which it is produced. We can also ask whether the processes of conceptual reduction and core categorization are compatible with conceptual richness and diversity characteristic of naturalistic inquiry.

The introduction of computer-assisted methods of analyzing qualitative data has encouraged claims concerning the theoretical power and analytic rigor in qualitative research. It is not surprising that developers of software for qualitative analysis have looked for (and found) inspiration in the path-breaking efforts of grounded theory to combine the richness of naturalistic inquiry with the rigorous canons of variable analysis. However, it seems that some of the key contributions of grounded theory to achieving this combination cannot be taken for granted. In the chapters which follow, I explore in more detail some of the issues raised by these efforts both to generate and to ground theory through qualitative research.

CHAPTER 3

Categories

Considering the key roles attributed to categories and coding in grounded theory, they have received surprisingly little systematic consideration in the literature on grounded theory (or, indeed, in other forms of qualitative inquiry). This may strike some as a very odd claim to make, given that Strauss and Corbin (1990) devote separate chapters in their introductory text to various forms of coding—open coding, axiel coding and selective coding. In their efforts at a fuller formulation of the procedures of grounded theory, these authors certainly place coding at the heart of the analytic process. The distinctions between open, axiel, and selective coding differentiate phases (or, better, emphases) in the process of coding data. However, they do not identify the key features of either categories or coding as analytic tools or distinguish these from other modes of qualitative analysis.

In this chapter, I shall try to identify (and clarify) the role that categories play in generating grounded theory. This requires a review of how categories are conceived in conjunction with and distinction from other key conceptual tools such as "properties" and "dimensions." I try to place the conceptualization of these terms on a surer footing, with regard to the analytic processes which they involve. In particular I distinguish the role of categories in formal classification from the centrality of properties in the substantive analysis of interaction. One of the strengths of grounded theory is that it embraces both classificatory and interactive forms of analysis. However, it is less successful in distinguishing these different analytic modes and methods.

As well as considering how categories relate to other concepts, notably properties and dimensions, we also have to consider their relationship with data. Here I look at Glaser's attempt to explain this relationship using a "concept-indicator" model. In reviewing the role of categories in grounded theory, I emphasize the need to consider the underlying analytic processes which this implies. Having clarified (I hope) the nature of categories and their general role in grounded theory, in the following chapter I turn to the use made of categories in the process of categorization.

CATEGORIES AS CONCEPTS

What is a "category" in grounded theory?

In the original text, categories are defined as "conceptual elements of a theory (Glaser & Strauss, 1967, p.36). This says little—but it implies a lot. In the first place, as elements in a theory, categories are "conceptual." This means that categories in grounded theory are more than simply names or labels which we attach to things in order to identify them. A name or label can be a concept, but it need not be one.

For example, suppose I assign fictional names to cases to protect the anonymity of my subjects. These names have no conceptual power or function beyond identification. The assignation of a fictional name (or label) tells me nothing more about the case. When I attach a name to a case, I merely designate the name to stand for that case. Designating involves specifying the name for a particular case, without implying any comparison with other cases—or anything about the case itself. It does not allow me to make comparisons between cases because it does not describe or characterize the case in any way. It is nothing more than an arbitrary string of symbols that serves to identify the case uniquely.

Of course, names and labels can have a conceptual import. The name "Elvis," for example, would be hard to disentangle from certain associations, positive or negative. And if I attach the label "strawberry jam" to a jar of jam, I am doing more than designating the jar of jam uniquely. I am assigning the jar of jam to a class (or set) of objects which are jams, and within that class of objects, I am assigning it to the subset strawberry jams. Indeed the label does not designate the jar of jam uniquely—it could easily be confused with any other (similarly labeled) jar of strawberry jam. But it does serve a conceptual function in comparing the contents of this jar with others (it contains jam) and of this jam with others (it is strawberry).

Even though the names and labels may sometimes serve the same conceptual function as categories, they should not be equated with them. This is because categories are always and inescapably conceptual, while names and labels may or may not be so.

> Category: *Class or division; one of a set of classes among which things are distributed.*
> Name: *Word by which an individual person, animal, place, or thing is spoken of or to; OR word denoting any object of thought, especially one applicable to many individuals.*
> Label: *Something attached to an object to indicate its name OR nature.*

Thus a category must be a concept and never just a name or a label. So far so good. But what kind of concepts are categories? According to the original definition, categories are concepts that "stand by themselves" as elements of a theory. This seems a strange claim—indeed, perhaps even contradictory. What does it mean to say that categories stand by themselves? And how can categories "stand by themselves" while at the same time being "elements of a theory"? Some light may be shed on this apparent contradiction by the distinction between categories and their properties.

CATEGORIES AND PROPERTIES

In contrast to categories, Glaser and Strauss suggest that properties cannot stand by themselves. Properties make sense only as aspects or elements of categories. To claim that categories stand by themselves is presumably to imply that categories are not conceived—or at any rate, need not be conceived—as the properties of other categories.

Unfortunately, in trying to illustrate this distinction, Glaser and Strauss offer an example that confuses rather than clarifies. They present "loss rationales" as a property of the category "social loss":

> . . . two categories of nursing care are the nurses" "professional composure" and their "perceptions of social loss" of a dying patient that is, their view of what degree of loss his death will be to his family and occupation. One property of the category of social loss is "loss rationales"—that is, the rationales nurses use to justify to themselves their perceptions of social loss. All three are interrelated: loss rationales arise among nurses to explain the death of a patient whom they see as a high social loss, and this relationship helps the nurses to maintain their professional composure when facing his death (Glaser & Strauss, 1967, p.36).

This illustration is rather puzzling, in at least two respects.

First, it is not at all clear what relationship is being illustrated between the category "social loss" and its property "loss rationales." In what sense is the latter a "property" of the former? The answer is not clear. The most obvious kind of relationship, that between a class and its members (that is, of inclusion and exclusion), does not seem to apply here. "Loss rationales" are not a subcategory of "social loss," it seems, but rather a separate category, referring to the strategies used by nurses in response to perceptions of social loss. These strategies may be related to their perceptions of social loss but the relationship does not seem to be one we might ordinarily describe as proprietorial—that is, that something is an attribute of something else.

Second, no clear distinction is drawn between the category relationships (how social loss relates to professional composure) and relationships between a category and its properties (social loss and loss rationales). The example in fact undermines the one plain hint we have been given—that categories can stand alone while properties cannot. The property of loss rationales is presented elsewhere as a concept that could be generalized to account for how all professionals distribute resources to clients. Although properties are not supposed to stand alone as elements of a theory, here a property does seem to stand by itself in a theoretical conceptualization.

For further discussion, let us turn to the way Strauss (1987) deals with categories in his solo effort to explicate grounded theory more fully. This explication was based on an "apprenticeship" model of learning and tried to convey some of the more subtle nuances of applying the methodology in practice. In his introductory account of categories, Strauss actually leaned very heavily on (and quoted extensively from) a text by Glaser that we consider below. But Strauss also introduced some new points to try explicate the relationships between categories, propertie,s and their dimensions. He provided "capsule definitions" of key terms [see inset box] suggesting that "it is essential to have a firm grasp of them" (Strauss, 1987, p.20). In a later text we are told that:

> Properties and dimensions are important to recognize and systematically develop because they form the basis for making relationships between categories and subcategories. And still later, between major categories. Therefore, to understand the nature of properties and dimensions and their relationships is a requisite task for understanding, in turn, all the analytic procedures for developing a grounded theory (Strauss & Corbin, 1990, p.70).

As these terms are presented as the building blocks of grounded theory, let us follow the advice Strauss offers and consider these definitions in some detail.

Dimensionalizing	A basic operation of making distinctions, whose products are dimensions and subdimensions
Category	Since any distinction comes from dimensionalizing, those distinctions will lead to categories.
Property	The most concrete feature of something that can be conceptualized.

The definition offered of a property as "the most concrete feature of something that can be conceptualized" suggests that Strauss considers properties to involve a lower level of abstraction than categories. Glaser (1992, p.38) also distinguishes between category and property in terms of levels of abstraction:

Category A type of concept. Usually used for a higher level of abstraction.

Property A type of concept that is a conceptual characteristic of a category, thus at a lesser level of abstraction than a category. A property is a concept of a concept.

Glaser here emphasizes the conceptual nature of both properties and categories and further suggests that a property is "a characteristic" of a category.

In the account presented by Strauss, matters are complicated by the addition of a third term, described as "dimensionalizing." Dimensionalizing is described as a "basic operation" that involves "making distinctions," though we are offered no account of how such distinctions are drawn. It is not quite clear what dimensionalizing adds to the process. Logically, it appears that we can equate categories and dimensions, since (1) dimensionalizing produces dimensions and (2) these involve distinctions that "lead to categories." In a later text, Strauss and Corbin themselves seem to equate properties and dimensions:

> Context: The specific set of properties that pertain to a phenomenon; that is, the locations of events or incidents pertaining to a phenomenon along a dimensional range (Strauss & Corbin, 1990, p.96).

Here properties are equated with points along a dimensional range.

To help clarify the nature of these distinctions, it may be helpful to consider the example offered by Strauss and Corbin. To illustrate their point, Strauss and Corbin discuss the properties and dimensions of the category "color." Color is described as having several properties: for example, shade, intensity, and hue. Each property has a dimension: intensity can be high or low, hue can be dark or light, and so on. So "locating along a dimensional range" refers to identifying the values (high, low, light, dark, etc.) of the various properties of a category when it appears in the data. If we are studying the color of flowers, we locate the color of any particular flower along a dimensional range by identifying the values of its shade, intensity, and hue. Flowers can then be grouped according to which values they share in order to explore relations between them.

This example comes closer to an orthodox distinction between properties and dimensions, in which something has a property which can vary along one or more dimensions—just as the space something occupies (its volume) can vary along three dimensions (height, width and depth). However, this may still seem rather confusing. We would normally think of the color of something (like the color of a flower) as a property that it has, and not a category that it belongs to. In contrast, color when considered as a category would not have properties, but subcategories—such as blue, yellow and so on. And we would normally think of shade, intensity and hue as the dimensions in terms of which each of these colours can vary, with "high" and "low" and so on as the units of measurement.

Let us therefore try to clarify the use of these terms. Strauss defines categories in terms of drawing distinctions, and this, at least, is familiar territory. The term category is commonly used to refer to a class of things—and the way we identify classes (and subdivisions within them) is commonly through the distinctions we draw between one group of things and another. These distinctions are commonly based on similarities and differences between things (though I shall qualify this assumption later). Categorizing therefore requires comparison.

The term property, on the other hand, is commonly used to refer to a quality or attribute of something, usually relating to an activity or accomplishment. Unlike dimensions (see below), properties often refer not so much to internal characteristics as to external relationships—how an entity acts on or interacts with its environment. Unlike categories, which relate things through classification, properties relate things through interaction. The point is aptly illustrated through the example of color, which varies according to physical nature and context of the interaction. Perception of color also depends to some extent (exactly how far is still under investigation) on the linguistic context, as some languages recognize a wider range of colors than others. This is a timely caution against assuming that any properties can be observed absolutely without taking account of both their natural and conceptual contexts.

The term dimension is commonly used to refer to extension—usually in terms of space or time, though we may talk of problems or ideas as having a number of different dimensions. Whereas belonging to a category implies a comparison, we can describe something (whether we refer to a thing, problem, or idea) as having a dimension (or a number of dimensions) without having to make any comparisons at all. Instead, we refer to some inherent (and measurable) extension—such as extension in space or over time. Identifying dimensions therefore involves (internal) differentiation rather than (external) comparison.

> Some Definitions
>
> Category: *Class, division, or any relatively fundamental concept.*
> Property: *Owning, being owned; attribute, quality, characteristic; quality common to a whole class but not necessary to distinguish it from others.*
> Dimension: *Measurable extent of any kind (e.g. length, breadth).*

Let us take water as another example. To categorize water, we can compare it with other things, such as alcohol or soft drinks. Or we could categorize it as a fluid rather than as a gas or a solid. Of course, considered as H_2O it can become either a gas (steam) or a solid (ice); but now we are comparing its chemical composition with that of other substances. Note that these categories are all assigned within a particular classification schema. When we distinguish water from alcohol, for example, we already subsume these under a superordinate category: that of beverages. Comparison is always for a purpose, and since our purposes vary, the distinctions we draw (and the categories we use) may also vary. Categories allow us both to identify what something is (identification) and to contrast it with other things (discrimination). Typically, we make these distinctions on the basis of the properties things have.

What properties does water have? Well, it can sustain life (we can drink it) but it can also destroy life (by drowning). These are only two of many such properties—water can also dissolve substances, expand on cooling, decompose on heating into hydrogen and oxygen, and so on. We can think of these as things that water can do in interaction with other things, or as things that we can do in interaction with water—heating, cooling, dissolving, drinking, and even drowning. Typically, then, we define properties through how something interacts with its environment, including ourselves.

Let us consider another example. If a thing has two legs, I am not likely to consider it as a tree (though trees sometimes have two trunks). The example is Harnad's (1987, p.539) and it is offered as an illustration of how we can try to distinguish trees from animals by means of a simple rule. If it has more than one leg, Harnad suggests then it is not a tree. The rule offers only an approximation and to deal with anomalies (and there are always anomalies) like a split trunk, we may have to improve upon it to get a better approximation. However, we may be misled by the emphasis in this example on passive observation (perhaps symptomatic of too much time spent in the laboratory). Properties are particularly useful as a basis for making distinctions because they involve interactions. If we want to distinguish trees from dogs, we do not need a series of successive approximations of what a tree looks like. Just whistle! Dogs can listen and come when you whistle. Trees cannot.

We might say that properties can be (but need not be) used as a basis for drawing (or inferring) distinctions between things—but it is the categories which express those distinctions.

Let us return to the example of water. What dimensions does water have? Physical dimensions are usually defined in terms of space (height, width, depth) and time. With water, we may measure its volume, though because it is fluid we will need a container rather than a ruler. Or we may measure its depth with a stick. Note that these dimensions are "internal" to the water we are measuring. They are not defined either in terms of interaction with the environment (at least, other than our measuring instrument) or through comparison with other things.

Some useful points about categories, properties, and dimensions emerge even from this very simple example.

First, we speak of something as having "properties" or "dimensions" but not of its having "categories." Thus water has dimensions and it has properties, but it does not "have" categories. Rather, we assign it to a category, depending on our purposes in making comparisons. Whereas properties and dimensions "belong" to the thing itself, the categories to which we assign it do not belong to the thing itself but are part of how we choose to classify it.

Second, we can speak of "discovering" properties and dimensions—by observing the way something interacts with its environment—but not of "discovering" categories. We may use these properties and dimensions as a basis for classification—but we do not observe or discover categories as such. Conversely, we can speak of creating or constructing categories but not of creating or constructing properties or dimensions.

Third, whenever we categorize, we make a distinction based on comparison—though we may not always make it explicit. But dimensions and properties need not involve such distinctions. We can identify dimensions (the depth of the water) or properties (its ability to dissolve substances) without making any comparisons at all. We are not claiming that this dissolves better than that, or this is deeper than that. On the other hand, when we categorize something ("this is water") we do make distinctions. We claim that it is one thing and not another. These claims relate to the cluster of attributes on which we have made our classification—some or all of which we may imply, depending on context. For example, if I hand over a glass with the assurance "it's only water" I may be implying that it is drinkable (or not, depending on the company I am keeping). If I encourage my children to swim in the sea with the assurance "it's only water" I may be implying that it will do them no harm. The properties I associate with the category vary with context.

Some Points about Categories

- *Things can have properties and dimensions, but they cannot have categories*
- *Categories involve distinctions based on comparison—unlike properties or dimensions, which need not involve comparison*
- *Categories can be created, but unlike properties or dimensions, they cannot be discovered*

Two further points are worth noting.

First, it is possible for us to make distinctions among properties and dimensions (that is, to categorize them) as well as among things themselves. We can draw distinctions about just about anything. For example, we can categorize the property of dissolving other substances as limited or extensive. Or we can categorize dimensions. We can measure the depth of a pond and compare it with the depth of the ocean; but it may be enough to know that the pond is shallow and the ocean is deep. Instead of measuring the depth of the water, we distinguish it in terms of the categories "deep" or "shallow." Any measure (such as depth or volume) can be simplified through categorization. But although we can categorize dimensions, we cannot dimensionalize categories. If we have not already measured the depth of the water more precisely, we cannot obtain that information from our categories "deep" and "shallow."

Second, we can use properties and dimensions as a basis for drawing distinctions—that is, as a basis for assigning categories. However, we can make distinctions (that is, categorize) on other grounds than the properties and dimensions that things have. Suppose we distinguish some subcategories of water that lies on land in different ways—puddles, ponds, streams, rivers, seas, and oceans. These distinctions may be based partly on a recognition of differences in some property—for example, water in puddles and ponds does not flow, while water in streams and rivers does. But puddles and ponds differ from streams and rivers in other ways too—differences that are not about properties at all, but are about how they are made, where they are located, when they dry up, and so on. I may distinguish the pond in the garden from the puddle on the pavement simply because I made one but not the other. So the distinctions that these subcategories express need not refer to the properties of water at all. Differences among properties may help us to categorize things, but we can do so on other grounds as well.

Further Points about Categories

- *We can categorize properties and dimensions, but categories do not have properties and dimensions*
- *Categories may be based on properties and dimensions, but they need not be*

Although they are related, the terms "categories," "properties," and dimensions" therefore refer to the rather different analytic processes. Let me emphasize this point by considering these processes in an area such as education.

One process involves classifying things into categories through comparison, involving relations of inclusion and exclusion. We can integrate categories (though superordinate categories) or refine them (through subcategories). For example, schools can be subcategorized in a variety of different ways, including: public and private; state and voluntary; primary, middle and secondary; free and fee-paying, single or co-educational; and so on (how we subcategorize obviously depends on our analytic purpose in making the distinction). This is how we use categories to make distinctions for the purposes of comparison.

A second process involves analyzing attributes or properties. For example, a school may have an enclosed yard for children to play in or playing fields for sports; it may have a distinguished history or a poor reputation; it may specialise in some subjects or offer a general curriculum; it may have an intake from all social classes or only one; and so on. Here, we are concerned with agency and interaction—crudely, with what schools do and how they do it. For example, knowing whether the school has an excellent teaching staff, a high standard of discipline, or an academic ethos may have a bearing on the kind of results the school obtains: whether it offers "added value" for a given intake.

A third process involves measurement along particular dimensions. For example, we could measure the size of schools in terms of the number of classrooms, the number of students, or the area the school occupies. Measurement is commonly associated with numbers but need not be. We may be quite content with "fuzzy" measures, such as high/low, big/small, many/few, and so on. Indeed, we may want to resolve fine-grained measurement (such as school income) into broader and more meaningful bands.

Three Analytic Processes

- *Categories can be assigned through comparison and classification*
- *Properties can be attributed through analyzing interaction*
- *Dimensions can be measured using a scale (which need not be numerical)*

Our confusion of these processes (in grounded theory and elsewhere) perhaps flows from the simple fact that we can apply all of them—assigning categories, analyzing properties or attributes, and measuring dimensions—to the same phenomenon. Take school fees, for example. We can use school fees as categories to distinguish between fee-paying and non-fee-paying schools; we can see them as properties or attributes of schools, which perhaps may have an effect (for good or ill) on admissions; and we can measure them along a dimension such as low to high. We may even want to subcategorize schools in terms of high or low fees to see what effects, if any, the size of fees have on admissions. It is easy to confuse the underlying processes when each can be applied to the same object. It does not help that categories may be inferred from differences among properties and dimensions, while properties and dimensions can themselves be categorized. Nevertheless, as I suggested earlier, each process has a different purpose. We use categories to distinguish and compare; we identify properties and attributes to analyze agency and effects; and we measure dimensions to identify more precisely the characteristics of what we are studying.

While the use of categories, properties, and dimensions in grounded theory apparently refers to these different analytic processes, it may be more helpful to distinguish them explciitly in these terms than to regard them as more or less interchangeable, or varying mainly their level of abstraction.

CATEGORIES AS ELEMENTS IN THEORY

So far we have considered how categories compare with properties and dimensions. Now we can consider their character as conceptual elements of theories. Although categories are said to stand by themselves they are also related through theory. What is the nature of this relationship?

In the original text, the relationship between categories seems to assume two aspects. The first aspect involves integration through higher levels of abstraction. It is suggested that "higher level, overriding and integrating, conceptualizations—and the properties that elaborate them" emerge in the later stages of analysis. The conceptual generalization from nurses to professionals may be regarded as an example of this process of abstraction. The second way of relating categories is through hypotheses, which involve the identification of "generalized relations" among categories. Glaser and Strauss illustrate these "generalized relations" in causal terms. For example, the following is offered as a hypothesis linking categories at the substantive level:

> The higher the social loss of a dying patient, (1) The better his care, (2) The more nurses develop loss rationales to explain away his death (Glaser & Strauss, 1967, p.42).

Although this hypothesis asserts two correlations, it does imply that generalized relations can be causal relations. These generalized relations can be identified, it is suggested, through the analysis of patterns of similarity and difference, which bring out "underlying uniformities" in the data. On the one hand, then, categories can be causally related, while on the other, these relations can be established through similarity and difference.

FORMAL AND SUBSTANTIVE RELATIONS

To avoid confusion, it may be helpful to consider further the relationship between these two different ways of relating categories. On the one hand, we have relations of similarity and difference, and on the other, we have relations of connection and interaction. The former is the familiar face of classification, where we lump things together that are similar and put them apart if they are not alike; for example, all public schools are similar in being public and they all differ from private schools in that respect. The latter involves the less familiar relations of connection and interaction, where our concern is with how one thing affects another.

A classic example of this difference can be found in evidence about the way children categorize. At the age of three, children presented with pictures of a dog, a cat, and a bone tend to group the dog and the bone—because the dog will eat (or bury) the bone. Older children tend to group the dog and the cat—because they are similar. The first distinction recognizes the interaction between dog and bone. The second distinction recognizes an abstract taxonomy ("animals") with which to group things in terms of similarity or difference. Sayer (1992, p.88) describes relations of similarity and difference as formal relations (cf. figure 3.1) and interactive relations between things as substantive relations (cf. figure 3.2). He adds that "although this is a very simple distinction, many approaches in social science have difficulty in recognizing relations of connection" (Sayer 1992, p.88).

One reason for this difficulty may stem from the problems of trying to identify substantive relations through formal classification. There are no (obvious) relations of similarity (or difference) between the dog and the bone. No doubt we could conjure some up—for example, dogs also have bones—but that is beside the point. Formal relations do not matter in understanding the interaction between the two. Our understanding of the substantive connection—the dog eats or buries the bone—is based on our observation of this process, as well as our own (bodily) understanding of hunger and how to satisfy it. We can understand that connection very well before we recognize that dogs are animals or

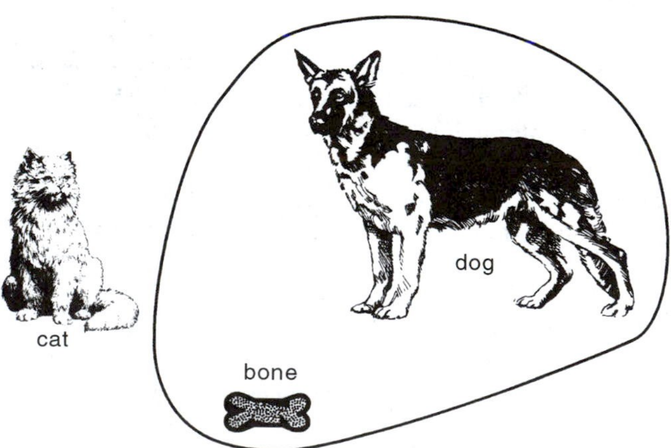

FIGURE 3.1 Substantive relations based on connection.

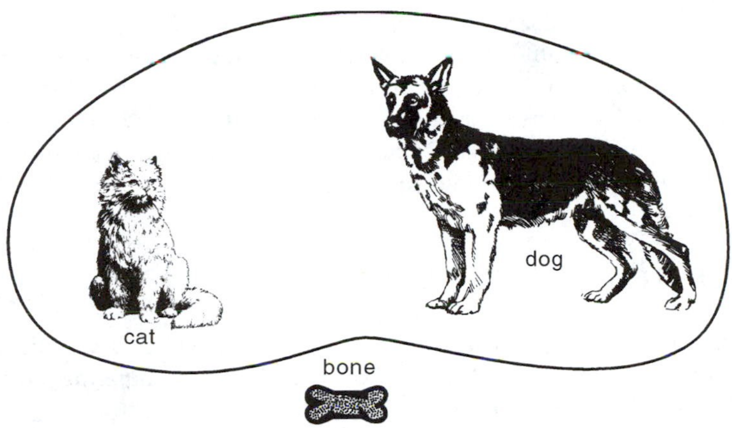

FIGURE 3.2 Formal relations based on similarity or difference.

that bones can be regarded as a form of food (for dogs). But we cannot understand that connection on the basis of formal relations because the things that interact—the dog and the bone—have nothing in common.

At this point, it might be objected that we can relate dogs and bones precisely because they have something in common—that they can be observed to "co-occur" on a regular basis. It might be argued that it is this regular co-occurrence that provides (more or less adequate) grounds for inferring a connection between dog and bone or, better still (from a theoretical standpoint), between dogs and bones. If this were so, then we should first classify our data and then look for connections among categories—just as sometimes seems to be suggested in grounded theory. But if we can observe (and experience) substantive relations directly (through our basic knowledge of the world)—as Glaser and Strauss also seem to argue on occasion—then this seems a rather convoluted procedure. Moreover, an emphasis on categorizing in terms of similarities and differences may obscure the very observations we could make in the first place about substantive relations between things.

Categories as Elements in Theory

- *Categories can be related in formal classification schemes based on similarity and difference*
- *Categories can be related in theories depicting substantive connections between them*
- *Substantive connections between categories can be identified directly and need not be inferred from relations of similarity and difference*

So far we have considered the role of categories with theory and how categories are related to other theoretical elements (such as properties) or our general theoretical aims (such as classification). Now let us consider how grounded theory presents the relationship of categories to the data we are trying to conceptualize. To do so, we need to turn to Glaser's account of categories in terms of a concept-indicator model.

CONCEPTS AND INDICATORS

According to Glaser, the concept-indicator model "provides the essential link between data and concept" (Glaser, 1978, p.62) through comparing indicators to each other and to the emerging concept:

> From comparisons of indicator to indicator the analyst is forced into confronting similarities, differences and degrees of consistency of meaning between indicators which generates an underlying uniformity which in turn results in a coded category and the beginning of properties of it. From the comparisons of further indicators to the conceptual codes, the code is sharpened to achieve its best fit while further properties are generated until the code is verified and saturated (Glaser, 1978, p.62).

The meaning of the category (or code—Glaser uses the terms interchangeably) is therefore defined in terms of its indicators. For Glaser, this does not require any quantitative analysis of the distribution or weight of indicators. Rather, it requires a conceptual analysis of their meaning through comparison which "slowly builds up to a concept and its properties" (Glaser, 1978, p.63). In this way concepts emerge from the data through the careful combination of indicators—and the distinctions they draw are determined by data.

Glaser also claims that this procedure prevents excessive "fracturing" of a concept through conceptual overelaboration into too many dimensions. Overelaboration, he suggests, derives from the uncritical use of "received" distinctions or the logical elaboration of concepts regardless of their empirical base. In contrast, only those distinctions that are "determined" by the data are conceptualized in grounded theory. This, he suggests, "helps achieve the goal for theory of parsimony of concepts, while at the same time richly densifying the theory" (Glaser, 1978, p.64).

The concept-indicator model of meaning is common in quantitative research, where it is conventional to define a concept in terms of its indicators. This allows abstract concepts to be "operationalized" for the purposes of research. For example, we cannot observe an abstract concept such as "prejudice" directly, but we can develop indicators of it—such as a set of statements with which the respondent can agree or disagree—that allow us to infer the existence of prejudice or to assess its degree. The meaning of the concept may be regarded as no more (or less) than the indicators that specify it in observational terms. Or the concept may be regarded as retaining a wider meaning which can be inferred from but cannot be reduced to the set of indicators through which it is observed.

Surprisingly, perhaps, given the "sensitizing" role envisaged for concepts, Glaser opts for the former position in which the concept is reduced to the set of indicators through which it can be observed. The concept is regarded as no more than the "sum" of its indicators—though no sums as such are involved, for it is the meanings of the indicators themselves that "count." Standard definitions of concepts are dismissed as "too restrictive"—it is the "operational meaning of the concept" that matters. Thus "conceptual specification is the focus of grounded theory, not conceptual definition" (Glaser, 1978, p.64). For example, the concept of social loss is not defined abstractly but specified in terms of particular ways in which nurses respond to patients—such as calculating the social value of patients.

On the other hand, indicators for a concept are also regarded as "interchangeable," so long as they can reasonably be related to that concept. Glaser gives the example of indicators of social value that involve comparison of different situations, such as the treatment received by various client groups (poor blacks, young mothers, derelict alcoholics and distinguished politicians) in different settings—including emergency rooms, intensive care units, cancer wards, and private rooms. The interchangeability of such indicators brings out "different values" so "greatly enriching the emerging theory" (Glaser, 1978, p.43). It allows the same concept to be explored in a variety of processes or settings which can bring out its different "dimensions." Any conventional concerns about consistency in the use of indicators are dismissed out of hand, since "the analyst is collecting facts as indicators to be compared and coded into ideas" and not "facts to be rendered empirically in descriptions" (Glaser, 1978, p.43).

Glaser's Concept-Indicator Approach

- *Categories are no more than the set of indicators that specify them*
- *Categories are not defined abstractly*
- *Indicators are related conceptually rather than numerically*
- *Indicators can be interchangeable across different settings*

Glaser's account emphasizes the generation of ideas through comparison of a variety of indicators in a range of contexts. Nevertheless, his use of the concept-indicator model raises some further questions about the role of categories in grounded theory.

One problem with specifying categories through their various indicators rather than defining them in terms of some consistent criteria is the loss of stable concepts that provide coherence within a theoretical framework. Glaser quite explicitly rejects the idea of stable concepts in favor of flexibility—if the concept is adapted to mean different things in different contexts, so much the better. However, if meaning is fragmented across a range of processes and settings, the relevance gained in immediate application may be at the expense of coherence in generalization (cf. Hammersley, 1989, p.122). Flexible concepts acquire their meaning(s) only within a specified local context. It is difficult to see how such a process can contribute to the parsimony in conceptualization that Glaser seeks.

A related problem lies in reduction of meaning to the indicators used in the research. This may undermine the potential role of concepts in theory. The conventional critique of reduction rejects the loss of meaning that follows when we reduce an idea to how it can be measured. Measurement (or observation) rarely captures the full significance of our ideas and, indeed, the most important aspects of meaning may be precisely those that resist any straightforward measurement or observation. Thus there may be more to categories such as "social loss" or "loss rationales" than indicated in the particular ways in which nurses deal with dying patients. Indeed, that conceptual power may be central to the process of generalization (to theory), which in effect abstracts these concepts from their particular "indications."

Reduction does have the merit of avoiding definition by fiat—that is, defining a concept in advance of analysis and then looking for data that "fit" the concept. Glaser is especially keen that conceptualizations "emerge" from the data. On the other hand, conceptual reduction seems inconsistent with the argument that categories once indicated by data acquire a life of their own, independent of the evidence through which they were generated. This implies that the meaning of concepts is not reduced to a set of indicators. This may seem essential if concepts are to play a more general role in with other concepts and integrating theory. But this also implies that concepts must be defined and not just empirically specified.

Some Problems with Glaser's Concept-Indicator Approach

- *If categories are not defined consistently, how is generalization possible?*
- *If categories are reduced to indicators, do they lose their theoretical power?*
- *If categories are reduced to indicators, how can they acquire "a life of their own"?*

Fortunately, we can reject conceptualization by fiat (which seems to be Glaser's overriding concern) without rejecting conceptualization per se.

Take, for example, the logical elaboration of categories. Glaser rejects this because it can produce "empty" categories that have no empirical reference in the data. He assumes that this will guide the researcher "into the realm of verifying what is supposed to be" instead of confining the research to what could be or is (Glaser, 1978, p.41). This assumption, however, may be false. The researcher may inquire about what logically could be the case without making any suppositions about it actually being the case. Asking a question need not presuppose a particular answer (though it may indeed presuppose a particular kind of answer). Logical elaboration may therefore be used to explore aspects of the data that might otherwise be overlooked. It allows the researcher to take note of what is missing in the data, as well as what is there. We can avoid a false consequence (that we devise categories and then look for the data to fit) without also rejecting the premise (that the logical elaboration of categories may pose some useful questions about the data).

IN CONCLUSION

We have established that categories are conceptual: that they are creative constructs, which express relations of similarity and difference but can also be connected theoretically through identifying substantive (causal) connections between them.

One of the distinctive aspects of grounded theory is its firm location in an interactionist methodology. Hence grounded theory is oriented to explicating "basic social processes" in dynamic terms—or, to put it crudely, how actions have consequences. The key methodology adopted to realize this ambition involves "constant comparison" across a range of settings where actors are engaged in the basic social process under investigation. This combination leads Glaser and Strauss to present the contribution of categories to theorizing in a rather ambivalent way. This ambivalence allows categories simultaneously to feature as the key conceptual tool in classification and in interactive analysis. However, no clear distinction is drawn between these two analytic modes, resulting in some confusion over the role of categories in generating theory.

I have tried to clarify this role by distinguishing between categories as a tool of classification and properties, which can be identified as an aspect of the analysis of interaction. As later versions of grounded theory throw in "dimensional analysis" for good measure, I have also tried to distinguish the analytic process involved in analyzing dimensions. The separation of these different modes of analysis—of categories, properties, and dimensions—may help researchers to steer a steadier path through the complications of a methodology which tries to unite different modes of inquiry.

This task may be aided by the further distinction between classification and causal analysis. These are better understood as two quite distinct modes of analysis with whose aims and objects are significantly different. It may be difficult if not impossible to base causal analysis on the classification—that is, on the identification of similarity and difference. I discuss this point more fully when we look later at how process is analyzed in grounded theory.

In relation to data, I have questioned Glaser's reduction of categories to indicators. I have suggested that this may undermine the conceptual role of categories in developing theory. This is a point which is explored in the next chapter, when I turn to how categories can be created. Our focus switches from categories as "things" (elements in a theory) to categorization as a "process." Given the key roles which categories play in grounded theory, we need to ask a simple question, which is surprisingly awkward to answer: namely, how the process of categorization can be accomplished.

CHAPTER 4

Categorization

In keeping with its roots in naturalistic inquiry, categorization in grounded theory is not seen as problematic. It is assumed that we all know how to distinguish and classify things—since, after all, we do it in everyday life. Analysis merely mirrors the "natural" process of attributing meaning and interpretation in ordinary interaction. However, this dependence on a taken-for-granted common sense may seem a rather insufficient basis for grounding a process so central to the generation of theory. It may therefore be helpful to reflect critically upon the basic processes in terms of which categories are constructed. Fortunately, as we shall see, the tools for such reflection are at hand in the form of studies of categorization in linguistics and cognitive psychology. These studies prompt us to look more reflectively at some of the issues concerning the way that categorization is used in the production of knowledge.

I do not mean to discuss these studies in detail so much as to identify and explore the light they shed on processes of categorization and the implications this may have for how we categorize in social research. Although categorization has become a fruitful field of inquiry in both cognitive psychology and linguistics, I do no more than select some of the main issues which seem pertinent in the current context. I pay particular attention to the work of Lakoff, who attempts to explain some of the new findings on categorization in terms of what he calls "idealized cognitive models." This approach if anything reinforces and deepens the emphasis in grounded theory on the conceptual nature of categories, while it also amends and extends in significant ways the classic model of categorization.

As we explore these studies of categorization, we will be taken further and further beyond the simple concept-indicator model that Glaser uses to account for the role of categories in grounded theory. The discussion identifies a number of processes involved in categorization beyond the judgments about similarity and difference that figures so largely in grounded theory. These analytic processes, I suggest, can provide a better basis for grounding categorization in grounded theory than the classic model to which Glaser appeals.

The concept-indicator model reflects the classic view of how meanings are identified through categorization. It is couched in terms of constant comparison between indicators (and indicators and emerging concepts) to identify similarities and differences, bringing out underlying uniformities. As we have seen, the concept-indicator model assumes that the meaning of a concept can be conveyed through indicators, that may express the meaning of the concept entirely or at least capture its essence. Meanings are treated rather like objects which can be reduced to (or assembled from) their constituent parts. This model has had a prominent place in social research, and its adoption in grounded theory has had an important role in extending its influence to qualitative inquiry.

Since grounded theory first originated, however, subsequent research on the use of categories has cast doubt on the adequacy of the concept-indicator model as an account of how categories are generated and used. The identification of categories on the basis of similarity and difference turns out to be rather problematic. The concept-indicator model assumes that things can be assigned unambiguously to appropriate categories—that we can establish clear rules governing what can or cannot be assigned to a particular category. However, in practice the process of drawing distinctions is much more complicated and ambiguous than the concept-indicator model allows. The concept-indicator model implies that categories can be defined in terms of their correspondence to features that we can observe. But, as we shall see, even within a rule-based approach to categorization, based on "invariant" observations, judgments about category membership are at best approximate.

APPROXIMATION IN RULE-BASED CATEGORIZATION

Let us begin with a "rule-based" approach to categorization that in some respects comes closest to the classical model. Even here we have to abandon any pretension to produce a clear and comprehensive basis for categorization.

Thus Harnad (1987) claims that categories are acquired through the identification of positive and negative instances among a range of "confusable alternatives." The features through which we make such identifications are never absolute—they are always approximate and provisional, and relative to the particular context of confusable alternatives encountered to date. Categories "converge" on an adequate representation, but the possibility of revision is always inherent given the prospect of new encounters, as we become acquainted with new or as yet undiscovered features of the object or situation.

Harnad suggests that categories can achieve some stability through the processes of "internal representation" involved in this learning by acquaintance. Such representation occurs at the perceptual and the symbolic level. At the perceptual level, iconic representation "preserves" the spatiotemporal shape or structure of sensory inputs through an image which abstracts from various features of an object. Such images are "unbounded"—they may reflect natural variations in sensory input, but at this level there are no clear divisions into different categories. These divisions occur through categorical representation, whereby iconic images are "labeled" and "sorted" in terms of their (putative) "invariant" features relative to alternatives with which they are confusable.

As an example, Harnad (1987, pp.551–2) offers the perception and categorization of a mushroom. The iconic representation of a mushroom preserves features of its perceptual "shape"—size, form, color, location and so on. The categorical representation involves eliminating (or "filtering") most of this "raw configural structure" to focus only on those features that help to resolve potential confusion (we don"t want to eat a toadstool). It selects (or abstracts) those features which provide a more reliable basis for judgment about category membership. The iconic representation offers a basis for the "relative discrimination" among sensory inputs—distinguishing the mushroom from the soil in which it grows, for example. The categorical representation provides a basis for what Harnad calls "absolute discrimination" for identification of the object as a mushroom (or even a particular variety of mushroom) and not something else (e.g., a toadstool) with which it might be confused.

At a further level of abstraction, symbolic representation involves the use of "invariance filters" which are derived from language rather than experience. As Harnad observes, "descriptions spare us the need for laborious learning by direct acquaintance" but he also argues that they depend on "the prior existence of a repertoire of labeled categories" on which descriptions can draw: "the descriptive system as a whole must accordingly be grounded in the acquaintance system" (1987, p.556). This qualifies the capacity of language to dictate experience, for no matter how abstract the descriptions which language can generate through recombining symbols, Harnad argues, they must ultimately be "grounded" in categorical representations. As we have seen, these categorical representations in turn filter features from shape-preserving iconic representations. Although flexible, our symbolic systems nevertheless depend on categories which can only approximate to and cumulatively converge upon our underlying experience.

Harnad's (putative) model of categorical representation adopts a "bottom-up" approach to cognition, suggesting that categories are "grounded" in experience, though not in ways which imply a one-to-one correspondence. To this end, Harnad sees categorization as a process of convergence on approximations of "invariant" features within a given context of confusable alternatives. This is a "rule-based" model of categorization, in which we devise better and better rules for identifying category members. According to Harnad, even such a rule-based model based on "invariants" supplied by experience cannot sustain "some absolute or ontic standard of veridicality" (1987, p.558), for meanings are identified in context and they are perpetually subject to holistic revision and updating in the light of further experience. The best we can achieve is convergence and a "steady tightening of approximation" guaranteed by "the cumulativity of the category formation process itself."

Harnad's Rule-Based Model of Categorization

Categorization provides rules to discriminate among confusable instances

Categories achieve stability through both iconic, categorical, and symbolic forms of representation

Categorical representation converges on invariant features

These "invariant" features are always approximate, context- dependent, and subject to revision

There is even less room for a "correspondence" view of categories where categorization proceeds in terms of idealized prototypes rather than definitional rules.

PROTOTYPICAL COMPARISONS

The classic experiment showing that we do not allocate items unambiguously to categories was performed by Eleanor Rosch in 1973. She asked students to rate how well as a number of items typified a category (Table 4.1).

Rosch found that students had no difficulty in rating items in terms of typicality on a scale of 1 to 7 (scoring a 1 for excellent). It took them on average only three seconds to rate each item, with a large measure of agreement over the ratings, especially regarding the best examples of each category (Table 4.2).

According to McNeill and Freiberger, this experiment took only 10 min-

TABLE 4.1 Items Used in Rating the Typicality of Category Members

Bird	Vehicle	Crime
Bat	Boat	Assault
Chicken	Car	Blackmail
Eagle	Horse	Embezzling
Ostrich	Scooter	Murder
Robin	Skis	Stealing
Wren	Tricycle	Vagrancy

Source: McNeill and Freiberger (1994, pp.84–5).

utes to perform but the results had "an impact which society has yet to absorb" as it "overturned the ancient tradition of thinking about human categories" (1994, pp.84–5). The experiment Rosch conducted showed (so they claimed) that assignment to categories is uncertain. It could no longer be assumed that categories have sharp boundaries so that something either does or does not belong to a category—for example, that something is or is not a "chair." Is the half-eaten apple still an apple? If it is drizzling outside, is it raining or not raining? Category boundaries become "fuzzy." Also, it could no longer be assumed that all things assigned to a category belonged to it equally—that everything defined as a "chair" is equally a "chair." For there are different degrees of belonging.

It is as though nouns function as adjectives, which can be more or less appropriate descriptions. Something can be rated as "more or less" a bird or a vehicle or a crime, just as we might rate them "more or less" large or small. We enter a strange universe here, where the seemingly solid world of everyday objects—even fruit and vegetables—dissolves into a flux of "fuzzy" categorizations. Nor is this fuzziness confined to settling border issues about category membership. It is not merely a question of some marginal cases, that we might be prepared quite happily ignore for everyday purposes. For ambiguity penetrates to the very heart of categorization, including the most familiar and conventional categories. Even those items apparently central to a class, such

TABLE 4.2 Rating Given for Typicality of Category Members

Bird	Vehicle	Crime
Robin (1.1)	Car (1.0)	Murder (1.0)
Eagle (1.2)	Scooter (2.5)	Stealing (1.3)
Wren (1.4)	Boat (2.7)	Assault (1.4)
Ostrich (3.3)	Tricycle (3.5)	Blackmail (1.7)
Chicken (3.8)	Skis (5.7)	Embezzling (1.8)
Bat (5.8)	Horse (5.9)	Vagrancy (5.3)

Source: McNeill and Freiberger (1994, pp.84–5).

as "robins and eagles," can still be differentiated as "more or less" category members in terms of their degree of typicality.

The evidence suggests that categories may be defined not so much in terms of common attributes (or indicators) as identified through their most typical examples. The best examples of a category can be called "prototypes" of that category—for example, a "robin" might be a prototype of the category "bird." A prototype is seen as an excellent example of a category because it shares many overlapping features with other members. But the features that make it exemplary (e.g. a robin is a wild bird) may be different from those which first qualify it as an ordinary member of the category (e.g., birds have wings). Indeed this research suggests that "the definition of a class yields few clues about its best members (McNeill & Freiberger, 1994, p.86) for in practice members of a category tend to share few attributes overall.

Membership of a category may therefore be assigned in terms of degrees of family resemblance to a prototype rather than through all members sharing some set of common features. Indeed there may be no features common to all members of the category, just as Wittgenstein found that the category "game" could be applied to various activities that bear a family resemblance but have no features common to all. Rosch suggests that prototypes allow categories to be represented (and understood) in terms of highly correlated clusters of attributes, thereby sparing us the need to worry about boundary definitions. Indeed, she argues that prototypes allow us to deal with categories on the basis of clear cases "in the total absence of information about boundaries" (Rosch, 1978, p.36).

Identifying Category Members

- *Membership of categories can be determined by kinship to an ideal or central (i.e., prototypical) member rather than conformity with a rule about attributes held in common*
- *Category membership becomes a matter of degree, not a dichotomy*
- *Category members need not share any common attributes*
- *Even prototypical members of a category may vary in their degree of typicality*

Illustrating with her experience the problems of initiating a paradigm shift, Rosch met very considerable opposition to her conclusion that category boundaries are illdefined. Her early papers were returned, and her audiences proved resistant. The idea that categories are "crisp" (McNeill & Freiberger, 1994, p.83) and can allow clear and complete conceptualization in terms of common features is rooted deep in Western thought. It reached its apogee in the program of the logical positivists to reduce ideas to their constituent elements, which were held to be observable. This program of specifying meanings in terms of observations proved intractable (and Wittgenstein, for one, abandoned it) but it nevertheless underlies the concept-indicator model of meaning in social research. The model assumes that categories can be defined "categorically" in terms of a set of common indicators. The identified set of indicators constitutes a necessary and sufficient condition for ascribing the category. But the prototype model of categorization certainly suggests that categorization is not quite so simple.

RULES, PROTOTYPES, EXEMPLARS, AND BOUNDARIES

While categorization may proceed on the basis of similarity and difference, the role of prototypes in categorization suggests that categories may be based on other procedures. Medin and Brasalou (1987) identify three methods of categorization and suggest a fourth.

The first does involve the identification of "rules" (which we might think of as indicators) to specify the common attributes required for membership. The rules governing category membership are stringent. A well defined category will have attributes that are jointly sufficient and singly necessary to identify the category. Only members of the category will possess all these attributes and all the members of the category will possess each one of them. For example, a "bachelor" may be defined as a never-married adult male. By this definition, anyone who is never-married, adult, and male can be categorized as a bachelor (the conditions are jointly sufficient) but no person who lacks one of these attributes (adult, male, or never-married) could be so categorized (each condition is necessary).

Sometimes membership can be determined less precisely by satisfying any of several conditions, each of which is sufficient, but none of which is necessary, to define membership. For example, Medin and Brasalou cite a strike at baseball, which occurs either when the hitter swings and misses/fouls or when the ball is in the strike zone. Another example might be the category "garage," which can apply equally to a place where you keep your car or a place where you fill it with fuel (or get it repaired). Nevertheless "the key characteristics of defined categories are that membership is all or none and that membership can be unambiguously determined by reference to a rule (Medin & Brasalou, 1987, p.461). Although this view is popularly perceived as the paramount method of categorization, Medin and Brasalou suggest that in practice it is less common than imagined. On reflection, it turns out that many categories are "fuzzy"—that is, "it is not possible to specify a rule that identifies all of its members and only its members" (Medin & Brasalou, 1987, p.461).

Medin and Brasalou also discuss the "prototype" approach to categorization. In their account, this involves identifying attributes that may be highly probable across category members, but they are neither sufficient nor necessary for categorization.

Suppose, for example, (as Medin and Brasalou do) that we include "eligible" in our list of attributes for bachelorhood. Now "eligibility" is less easily identified than adulthood, gender, or marital status—though each of these attributes can also give us problems, particularly the last (marital status), given the rapid rise in the numbers of adult men who cohabit. We may think of eligibility in terms of a stereotype: someone with a high income, good career prospects, and strong on sex appeal would probably do nicely. So we may consider a never-married adult male to be "eligible" insofar as they conform to some (but not necessarily all) of these characteristics. On the other hand, we also think of "bachelors" as men who are dedicated to their nonmarital status—far from being "eligible," they like to live life in the fast lane. Either way, our categorization involves identifying kinship with a prototype rather than assignation by strict definition.

Another method of categorizing identified by Medin and Brasalou involves comparison with exemplars that are drawn from memory (rather than abstracted from experience, like a rule or a prototype). If I have encountered an example of a category in previous experience, I can compare the new candidate with that previous example as a method of determining membership. For example, suppose we know very well a relative who is a "confirmed bachelor"—and then encounter a never-married male with a similar outlook on life. We might categorize this male as a bachelor, too, on the basis of our knowledge of the previous example. Here it is kinship with a previously encountered example rather than conformity to a rule that is used to determine category membership.

As well as categorizing through rules, prototypes and exemplars, Medin and Brasalou suggest a further basis of categorization, through boundary definitions. These may apply where attributes have continuous values and we have to judge where one characteristic shades off into another. For example, we might decide that only those males over a certain age should count as bachelors in our categorization of bachelorhood. Boundaries involve comparison at the edges of a category, rather than comparison with its central features or in terms of some ideal characteristics.

Thus there are different ways of categorizing, and even assigning as seemingly straightforward a category as "bachelor" may involve a variety of methods. This variation in categorization is important because the method adopted has implications for the ways that we use categories. For example, the possibility of error in categorization differs according to the method employed. There seems more prospect of identifying errors where categories are well defined and membership is governed by rules than where they are "fuzzy" and

The Methods of Categorization Identified by Medin and Brasalou (1987)

- *Rules involve definitions of attributes that are jointly sufficient and singly necessary for category membership*
- *Prototypes involve identification of attributes that are highly probable but not common to all*
- *Exemplars involve comparison with previous cases*
- *Boundary definitions involve comparison of attributes for discrimination at the margin*

require comparisons with prototypes or exemplars. If membership is graded and members can belong "more or less" to a category, it becomes even more difficult to be "categorical" about error.

THE AIMS OF CATEGORIZATION

The conventional emphasis on categorization based on rules that define similarity or difference probably stems in part from the common equation of categorization with classification. Although we tend to equate categorization with classification, it may be helpful to recognize that there are a variety of different purposes associated with categorization. Apart from classification for its own sake, Medin and Barsalou suggest three other uses:

- Inference and prediction
- Generation
- Productivity

Categorization is important for inference and prediction. If you categorize someone as a bachelor, this may have significant implications for how you relate to them—at least in some social contexts. If you are about to invest some effort in seduction, for example, you may want to avoid a category mistake which could end in wasted time and unwelcome embarrassment. To categorize someone as a bachelor, you want to identify attributes with high "cue validity"—which means there is a high probability that something with an attribute is a member of the category. To infer attributes from category membership, on the other hand, it is more useful to identify attributes with high "category validity." This means that there is a high probability that something has the attribute if it belongs to the category. The attributes that are good for diagnosis may be poor for inference. For example, every bachelor is an adult male, so these attributes (adult, male) have high category validity—once you know someone is a bachelor you can infer they are adult males. But adult male has low cue validity for being a bachelor, since many adult males are married.

Generation, like prediction, involves going from a category to exemplars, but in this case by finding or "instantiating" examples of the category. For illustration, Medin and Barsalou offer the example of a camping trip, where you can "instantiate" various examples of categories of clothing, food and so on that you need for the trip. These categories ("clothing for trip," "food for trip") are created off the cuff and testify to the very flexible and fluid character of categorization. The items that are instantiated may depend on just what food happens to be in the cupboard or which clothes have not been consigned to the dirty clothes basket. Categories and category members can be created and identified as required—though you may maintain a list of items needed for a holiday, organized under categories "clothes," "food", and so on to save reinventing the wheel each time you go on vacation.

By "productivity," Medin and Barsalou refer to the production of new categories, as categories are combined to produce more complex concepts. They give the example of the categories "disgusting," "obnoxious", and "flea" being combined to form the more complex concept of a "disgustingly obnoxious flea" (1987, p.466). The productivity here seems to inhere in adjectives that can be used to qualify other categories—so some categories may have higher "productivity" than others. The more general point is that categories can be used in combination so that we have to consider their relationships with other categories as well as with the objects that they are intended to categorize.

Uses of Categories (Other than Classification)

- *Finding category instances*
- *Predicting category membership (needs high cue validity)*
- *Inferring attributes from category membership (requires high category validity)*
- *Combining with other categories to produce complex concepts*

These various uses of categorization may explain why it may be more salient to establish kinship with prototypical members and previous exemplars or to discriminate at category boundaries than to devise formal rules governing membership. Apart from any other considerations, we need only consider the difficulties of establishing clear rules and the efficiency and practical reliability of the alternative approaches.

CATEGORY JUDGMENTS

How do people make judgments about category membership when they compare examples with prototypes or exemplars? Medin and Barsalou distinguish between "holistic processing" and "analytic processing" of categories, suggesting that sometimes people focus on their overall impression of the example. This is typically so where they are under time pressure to make a quick judgment or they are unable to isolate separate attributes.

Where people can identify separate attributes, how do they combine decisions about these to make an overall judgment? Medin and Barsalou distinguish here between "additive" and "multiplicative" judgments. The former is possible where the outcomes of each comparison do not affect the others—for example, where being male and being adult can be regarded as independent attributes. You can be male and not adult or adult and not male. But what if the outcomes of one judgment are connected with another? For example, suppose we think of age in terms of maturity, and our judgments about maturity are affected by gender considerations. We may be more willing to confer adult status on young women (because they mature more quickly) than young men. Here our judgment about adulthood may reflect our judgment about gender. Or suppose that age, marital status and eligibility are connected in some way. Perhaps an older man is seen as less eligible because he has never married or perhaps he may be seen as more eligible because he is more likely to have higher income. Then judgments about eligibility may reflect judgments about age. The reasoning involved in making such judgments becomes more complicated if the outcomes are interrelated.

The complexity and fluidity of category judgments suggests that the process of categorization is much more variable than the concept-indicator model allows. While we may assume that "it is essential that people's conceptual systems be similar and relatively stable" for the purposes of communication, research suggests that "categories are much less stable than would be expected, both between and within individuals" (Medin & Barsalou, 1987, p.468). While this lack of stability in category judgments provides some support for Glaser's acceptance of the "interchangeability" of categories, it also challenges any simple assumption that categories are "indicated" by data in a straightforward way.

Logical Connections

Classic classification assumes that the world exists in a particular way, made up of individual entities that have identifiable properties and relations. Each entity has some distinctive properties that are integral (or essential) to its nature, while other properties and relations may be incidental. Those entities that share essential features can then be categorized as belonging to the same class. Membership can be clearly assigned on the basis of rules or necessary and sufficient conditions. Relations between categories (and the entities they represent) can be defined in terms of formal logic. If category A contains B, then all entities categorized as B will share the same properties as A (all Bs are A). If category A does not contain B then no B can be an A.

More complex classifications can be formed on the basis of these primitive (or atomic) categories, through the logic of set theory, which allows the relationships between categories to be defined through logic (and mathematics) via intersections (A AND B), unions (A OR B) and complements (A NOT B). Classification proceeds from logical primitives to more complex forms, in terms of superordinate categories structured through hierarchical organization. Symbolic representation mirrors the natural world of individual entities with their distinctive properties and relations. Symbols (categories) therefore acquire meaning through their correspondence with entities, properties, and relations in the real world. The meaningfulness of a categorization then depends on its correspondence with this objective reality or, more precisely, on its "truth conditions"—that is, the conditions under which it would be true.

Once category boundaries become fuzzy and membership becomes a matter of degree, the logic that underpins classification is undermined. This logic is embodied in two "laws" that date back to Aristotle (McNeill & Freiberger, 1994, pp.52–3). First, the law of contradiction states that "A cannot be both B and not-B." Once we accept that being B is a matter of degree, then A can be both B (to some extent) and not B (to some extent). So I can have a friend (to some extent) who is not a friend (to some extent). The law of the excluded middle states that "A must be either B or not-B." Once we allow that A can be B to some degree, we must abandon the claim that it must either one thing (B) or the other (not-B). So I can have a friend "of sorts" and the world is no longer populated by only those who are either friends—or not. Once these laws go, truth claims become a question not of either/or but a matter of degree. Things become true only "to some extent."

The implications of accepting that category boundaries are fuzzy and membership of categories may be graded or based on family resemblance to exemplars or prototypical members may therefore be profound. However, there is little point in undermining the classic model, if we cannot produce an alternative account of how categorization can proceed. Let us therefore explore more fully the complicated ways in which category judgments can be made, making use of the seminal work by Lakoff (1987) on categorization.

IDEALIZED COGNITIVE MODELS

Reviewing the evidence of prototypical categorization, Lakoff suggests that category judgments are made in terms of what he calls "idealized cognitive models." These models are not accurate representations of the real world—they are "idealized." Nor are they representations of that world—they are irreducibly cognitive. Thus category judgments concern observations and attributes but are made within an underlying cognitive context that informs category judgments.

For example, Lakoff cites an argument (from Fillmore) that the category "bachelor" exists

> as a motivated device for categorizing people only in the context of a human society in which certain expectations about marriage and marriageable age obtain (Fillmore cited by Lakoff, 1987, p.70).

Fillmore argues that the category "bachelor" would not ordinarily be applied to cohabiting men, tarzans or popes. The reason, Lakoff suggests, is that the category "bachelor" is applied with respect to an "idealized cognitive model" of a human society "with (typically monogamous) marriage, and a typical marriageable age" (1987, p.70). In other words, a reliable category judgment can only be made to the extent that the idealized model fits the world we are trying to describe. This judgment is therefore "irreducibly cognitive," requiring the comparison of different cognitive models—such as those which characterize cohabitees, popes, and tarzans. The appropriateness of our judgment is not determined directly by correspondence with the world "out there." Instead of trying to establish a set of rules to establish correspondence, this suggests that category judgments depend on how these idealized cognitive models are generated and applied.

Making Connections

According to Lakoff, each idealized cognitive model can be regarded as a complex whole that may be structured in terms of different principles. These include:

- Propositional structures
- Image-schematic structures
- Metaphoric mappings
- Metonymic mappings

Much of our knowledge takes the form of propositional structures that "specify elements, their properties and the relationships holding among them (Lakoff, 1987, p.113). They are "network structures with labelled branches that can code propositional information" and "contain empty slots, which can be filled by the individuals occurring in a given situation that is to be structured" (Lakoff, 1987, p.116). The idealized cognitive model underlying the category "bachelor" is an example of such a propositional structure. This propositional structure relates to the rule-based approach to categorization, but in Lakoff's account, the rules can only be understood as parts of an overall cognitive structure.

Image-schematic structures shape categorization in accord with a specific image, typically rooted in perceptual or bodily experience. For example, the "container" schema defines a basic distinction between "in" and "out" which reflects both bodily experience (e.g., we breathe in and out) and everyday experience (e.g., we move in and out of rooms). The "container" image involves an interior, a boundary, and an exterior—elements that underpin the logic of classification. For example, the logical claim "If all As are Bs and X is an A, then X is a B" follows from our understanding that if A is contained by B and A contains X, then B contains X. However, Lakoff argues that container images are meaningful mainly by virtue of our bodily experience rather than by virtue of their logical power.

As well as the container image, Lakoff identifies other image schema that structure cognition. These include the part-whole schema, the center-periphery schema, and the source-path-goal schema, all of which involve differing structural elements (and logics) derived from bodily experience. Such images often structure the extension of categories beyond a basic core. For example, Lakoff considers the Japanese classification "hon," which is applied to objects which are long and thin and preferably rigid, such as sticks, canes and pencils. The word is also applied where this feature is not evident except through association. For example, martial arts (which use long thin rigid weapons) can be classed as "hon." So too can TV programs (which are like telephone calls but without the long thin wires). Thus the image schema becomes extended to objects or activities which bear no common features with the central objects.

Two common methods of structuring extensions are provided by metonyms and metaphors.

A metonymic extension involves the substitution of some attribute or adjunct for the thing itself. The martial art becomes classed in accordance with one aspect of the art—the use of long, thin, rigid weapons. This can even be extended to judo, which can also be classed as "hon" despite its lack of weapons (but it can be classed in accordance with its aspect as a martial art). Metonymy—where some aspect stands for the whole—is a major source of prototype effects in categorization (Lakoff, 1987, p.79).

A metaphorical extension involves a mapping from a model in one domain to a corresponding structure in another domain. For example, the container model within its in/out imagery derived from bodily experience can be "mapped" to other domains. We noted above the metaphor of classes as containers, with items "in" or "out" of a particular class. Lakoff finds the container metaphor mapped to many fields, including activity (we "overdo" it), anger (we become "agitated"), mind (we have "ideas" or we feel "out of it"), vision (things go "out of sight"), and personal relationships (which we get "into" or "out of").

Metonymy: *Substitution of the name of an attribute or adjunct for that of the thing meant.*
Metaphor: *Application of name or descriptive term or phrase to an object or action to which it is not literally applicable.*

Complexities of Categorization

Lakoff argues that these propositional and image-schematic structures and metonymic and metaphorical mappings can explain some of the complexities of categorization, including prototype effects. As we have seen, the category "bachelor" provides an example of prototype effects, as identification reflects an idealized cognitive model of marriage and marriageable age. This model may be more or less apt in a given context—we are more likely to find bachelors on the football field than in the jungle or the papacy. Thus the boundaries of the category reflect the purchase of the idealized cognitive model that informs it. But even if the cognitive model "fits" the context, there are prototype effects within the boundaries of the category that are associated with our stereotypes of what bachelors do. Stereotypes help to define normal expectations. We expect bachelors to be "macho, promiscuous and nondomestic" rather than to like children, prefer stable relationships and love housework (Lakoff, 1987, p.86). Those who conform to the stereotype are more likely to be seen as good examples of the category "bachelor" than those who do not. In Lakoff's view, this stereotypical categorization involves a metonymic substitution of part for whole.

The Complexities of Motherhood

Lakoff explores some of these complexities through the example of motherhood. This also involves a stereotypical categorization in which a subcategory "has a socially recognized status as standing for the category as a whole," in this case the "housewife-mother" subcategory (Lakoff, 1987, p.79). Thus the "working mother" is not just a mother who happens to work but is implicitly defined in contrast to the stereotypical housewife-mother. The underlying cognitive model here is of motherhood as "nurturance" rather than "biology"—so that working and unwed "mothers" whose children are adopted are not deemed "working mothers." To be a "working mother," you have to "mother." Though the stereotype "housewife-mother" stands for the whole category, it is defined relative to one of several cognitive models of "motherhood." Lakoff distinguishes a number of models of motherhood (Table 4.3).

TABLE 4.3 Idealized Cognitive Models of Motherhood

Model	Definition of motherhood
Birth	A woman who gives birth to a child
Genetic	A woman who contributes genetic material to a child
Nurturance	A woman who nurtures and raises a child
Marital	The wife of the father of a child
Genealogical	The closest female ancestor of a child

Source: Lakoff (1987).

The category "mother" therefore involves a complex model in which several individual models cluster together. However, Lakoff suggests that there are strong pressures to select one of these as primary, particularly when the individual models diverge. However, there may be no agreement about which model is paramount, even among those defining the term in dictionaries—where the birth, nurturance and genealogical models have all been given primacy. Thus there are no necessary and sufficient conditions that define motherhood once and for all, which are common to all mothers.

Lakoff argues that various subcategories of motherhood (stepmothers, adoptive mothers, surrogate mothers, etc.) are not defined in terms of common features but "by virtue of their relation to the ideal case, where the models converge" (1987, p.76). This ideal or central case involves:

> . . . a mother who is and always has been female, and who gave birth to the child, supplied her half of the child's genes, nurtured the child, is married to the father, is one generation older than the child, and is the child's legal guardian (Lakoff, 1987, p.83).

The various subcategories are deviations from the central case (though not all deviations may be recognized by subcategories). No general rule generates these subcategories, for they "are culturally defined and have to be learned (Lakoff, 1987, p.84). Indeed the ideal case of a "mother," as Lakoff presents it, is clearly a historical product, requiring some stipulations (always female, supplying genes) that have only acquired salience in recent years.

Radial Structures

Thus subcategories vary according to convention rather than logical derivation. This conventional variation (rather than logical derivation) of subcategories from a central case Lakoff describes as a "radial structure," which, he suggests, is "extremely common" in categorization. Although the extension of a central case to variants is not governed by general rules, he suggests that it is not arbitrary either, for the various cognitive models (propositional, im-

TABLE 4.4 Classification in Dyirbal

Category	Examplars
Bayi	Men, kangaroos, possums, bats, most snakes, most fishes, some birds, most insects, the moon, storms, rainbows, boomerangs, some spears, etc.
Balan	Women, bandicoots, dogs, platypus, echidna, some snakes, some fishes, most birds, fireflies, scorpions, crickets, the hair mary grub, anything connected with water or fire, sun and stars, shields, some spears, some trees, etc.
Balam	All edible fruit and the plants that bear them, tubers, ferns, honey, cigarettes, wine, and cake
Bala	Parts of the body, meat, bees, wind, yamsticks, some spears, most trees, grass, mud, stones, noises and language, etc.

Source: Dixon (1982) cited in Lakoff (1987, pp.92–93).

age-schematic, metaphoric, and metonymic) provide the principles for extension.

To illuminate the role of radial structures in classification, let us follow Lakoff on an excursion into the linguistic world of Dyirbal, an Australian aboriginal language. As another example of a radial structure, Lakoff discusses an account by Dixon (1982) of classification in traditional Dyirbal. The four words "bayi," "balan," "balam," and "bala" classify all the objects in the Dyirbal universe (Table 4.4).

There are many intriguing points raised by this classification. You may have noticed that "some spears" are included in three of the four categories—suggesting that any rules governing this classification are far from straightforward. Dixon identified some general principles guiding the assignation of categories (Table 4.5).

Lakoff notes the way these general principles are extended to items that do not share similar features. One extension is through sharing a "domain of experience." Thus fish are animate and are classed along with animals as "bayi"; and fishing spears are in the same domain of experience and so are also classed as "bayi." The general principles are also modified where special cases modify their application. One modifier Dixon proposed is an "important property" principle, whereby an item may be transferred to another class if be-

TABLE 4.5 Principles underlying classification in Dyirbal

Bayi	(human) males; animals
Balan	(human females; water; fire, fighting
Balam	nonflesh food
Bala	everything else

Source: Dixon (1982) cited in Lakoff (1987, p. 93).

longs to a subset with a distinctive characteristic, such as (most commonly) being harmful. Therefore some fish perceived as harmful are designated as "balan" rather than "bayi." Likewise the hairy mary grub, whose sting is said to feel like sunburn, is classed as "balan," though this may also reflect its association with fire. Another modifier is a "myth-and-belief" principle, whereby an item is classed according to its mythical connections rather than its own characteristics. Though birds are animate, they are generally believed to be spirits of dead human females and are classed as "balan." Some species of willy-wagtail, on the other hand, are believed to be mythical men and are classed as "bayi." Lakoff suggests that both these modifiers reflect the crucial role of domains of experience in shaping categorization systems.

Lakoff suggests that the Dyirbal schema illustrates the general principles underlying categorization. These can be summarized as follows:

Centrality	Some members are more central than others
Chaining	Central members are linked to other members
Experiential domains	Links are characterized by culturally specific domains of experience
Idealized models	Links are characterized by idealized models of the world
Specific knowledge	This overrides general knowledge
The other	There is a residual category for "everything else"

Categorization along these lines does not proceed on the basis of common characteristics. There is nothing common to "women, fire, and dangerous things" (which Lakoff takes as the title of his book) such that their membership of the "balan" category can be derived (or predicted) from a set of logical rules. The way categories are structured reflects culturally specific domains of experience (such as Dyirbal myths) that allows us to make sense of categorization (in Lakoff's terms, we can understand what "motivates" it) but not to predict its outcomes.

Nor does categorization proceed through the assignation of primitive elements that can be then used to create more complex categories through combination. As well as emphasizing the role of idealized cognitive models, Lakoff makes two further points against this view of categorization. One is that categorization begins with basic categories that already involve an internal structure. The other is that complex categories cannot be understood merely as combinations of more simple elements.

Basic-Level Categories

According to Lakoff, basic level categories are neither the most general nor the most specific: they are somewhere midway between. Superordinate cate-

TABLE 3.6 Examples of Different Category Levels

Superordinate	Basic	Subordinate
Animal	Dog	Retriever
Furniture	Chair	Rocker

Source: Lakoff (1987, p.46).

gories are generalizations "up" from basic level and subcategories are specializations (or refinements) "down" from the basic level.

The basic level may be seen as basic in a variety of senses. In terms of perception, basic-level categories are those with which we can apprehend members in a single mental image—they have a similar overall shape that can be readily identified (e.g., dogs and cats). In terms of function, basic-level categories are those of entities with which we commonly interact—we pick flowers, stroke cats, and play with the dog. In terms of communication, basic-level categories are the shortest, the most commonly used, and those we learn first (as children). In terms of organization, most attributes are identified at the basic level (dogs have paws, tails, and teeth, etc.).

Basic-Level Categories Have Members

- *We can apprehend in a single mental image*
- *With whom we commonly interact*
- *Whose attributes are also basic*

Lakoff contests any identification of basic level categories with "natural kinds" that are posited to exist independent of human perception (and action). These categories are not basic because they reflect a natural order which decrees that there are dogs, cats and chairs. They are basic because of the way we interact with the world, which is shaped by our general human capacities for perception and action (we are not flies or elephants).

> Basicness in categorization has to do with matters of human psychology: ease of perception, memory, learning, naming and use. Basicness of level has no objective status external to human beings. It is constant only to the extent that the relevant human capacities are utilized in the same way (Lakoff, 1987, p.38).

This permits the possibility of cultural variation in the identification of basic categories—urban dwellers may identify "tree" as a basic level category while rural dwellers recognize different genus (oaks, etc.). Lakoff suggests that our basic level knowledge is organized around gestalt perceptions of part-whole divisions, in which basic-level wholes are constituted through parts have significance for functions, shape, and interaction (for example, we sit on the seat of a chair). The "whole" here is more than the sum of the parts—just as the chair is more than the separate parts (legs, seat, and back) of which it is made—and it is also perceived as simpler than its individual parts. Thus basic-level categories tend to have an internal structure—but they are not simply combinations of parts. For example, the parts of a chair—the seat or the back—cannot be grasped separately from the concept of the chair itself.

Complex Categories

If basic level concepts have an internal structure, they cannot perform a function as atomic primitives (or indicators) in a logically structured classification scheme. In any case, Lakoff argues that complex forms of categorization cannot be understood as mere aggregations of simple categories. In classic accounts of categorization, complex categories can be identified by the "intersection" or "union" of primitive categories. Where two categories intersect, they create a new complex category, whose members then share the characteristics of both the primitive categories. For example, take "fat" and "friends." The intersection of these categories defines the complex category "fat friends" (Fig. 4.1).

Where two categories unite, they create a complex category whose members have the characteristics of either primitive category. For another example, take "relatives" and "friends" (Fig. 4.2). If we unite these categories, we create a complex category, that we might call "intimates." In pictorial terms, intersection is defined by the overlap between categories, while union is defined by their conjunction.

This account of complex categorization assumes that category boundaries are crisp rather than fuzzy—that we can decide with confidence about what

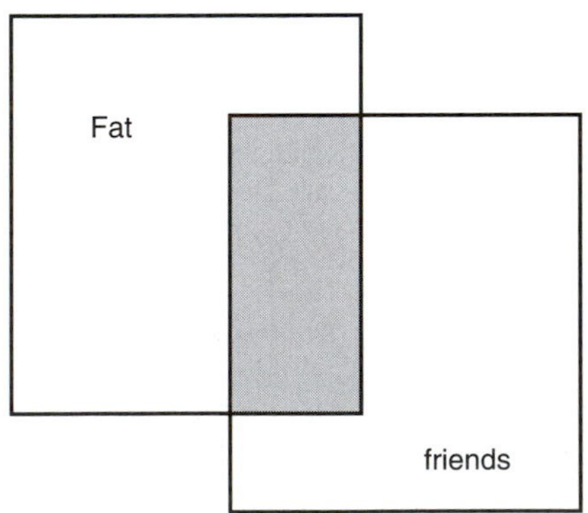

FIGURE 4.1 Intersection of categories.

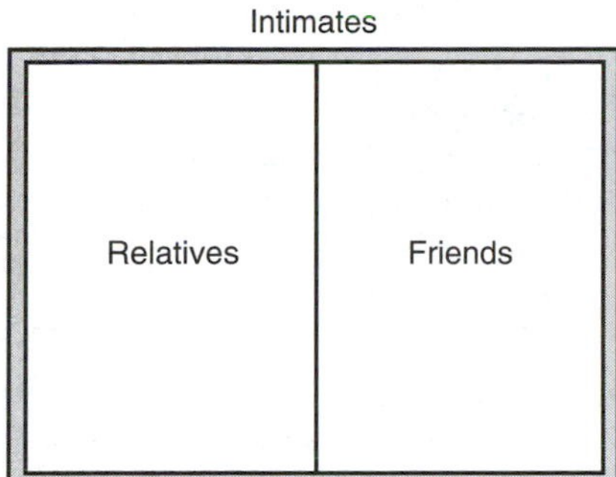

FIGURE 4.2 Union of categories.

belongs where. But what if we cannot be sure—because "fat" or "friendship" is a matter of degree or membership is determined by conformity to prototypes rather than clear rules or conditions?

Complex categories give rise to other problems too. Suppose we consider the category "faithless friends," which might be considered the result of the intersection of people who are friends and people who are faithless. Those in this category could not be cited as good examples of friends (who generally are not faithless) nor as good examples of faithless people (who are generally are not friends). "Faithless friends" are not easily perceived as a product of the intersection of these categories, any more than "small galaxies" can be seen as an intersection of things which are small and things which are galaxies.

Lakoff suggests instead that we understand complex categories such as "faithless friends (the example is mine but the argument is Lakoff's) through the creation of an idealized cognitive model for that category. This is possible because we can combine elements of the cognitive models which underlie each category—"faithlessness" and "friendship"—to constitute a new model of "faithless friends." These elements are not combined through some logical formula but rather "motivate" the development of a new cognitive model so that "the meaning of the whole is often motivated by the meanings of the parts, but not predictable from them" (Lakoff, 1987, p.148).

Motivation refers to the acquisition of meaning through family resemblance (Lakoff, 1987, p.65) along the lines of the structuring and mapping principles already discussed (presumably rather like we "invest" something with meaning). Lakoff contrasts motivation with knowledge governed by algorithms, which allow one to state rules to compute or predict an output from a given input. In algorithms, knowledge is either predictable or arbitrary, but motivation produces meanings that are "neither arbitrary nor predictable (Lakoff, 1987, p.346). Meanings (or categories) are "motivated" conventionally, through links, associations, and substitutions (e.g., metaphorical and metonymic extensions) that reduce arbitrariness and make them easier to learn and use.

Lakoff's approach also explains the problems of classical logic. In Lakoff's view, the difficulty stems from applying the logic of mathematics to real world examples, that do not conform to the mathematical ideal. In formal logic, "bachelors" are defined as "unmarried men" (all As are B) and it follows that all unmarried men are bachelors (all Bs are A). However, this logical equivalence breaks down once we consider bachelors and unmarried men as categories that are meaningful in terms of underlying cognitive models of "bachelorhood" and "unmarried men." As the underlying cognitive models differ, we cannot claim that "all unmarried men are bachelors" as this conflicts with some of the assumptions which motivate (make meaningful) the category "bachelor." If definitions are made "relative to cognitive models that are idealized, models that may or may not fit the world well and need not be consistent with one another" (Lakoff, 1987, p.130) then we can expect to find unmarried men (e.g., Tarzan, the Pope) who are not "bachelors."

IMPLICATIONS FOR GROUNDED THEORY

First I shall try to summarize some of the main points to emerge from this lengthy excursion into the intricacies of categorization. Then let us consider the implications for grounded theory.

In the above discussion, we can identify at least four different accounts of categorization. First we have the classic account, which assumes that category boundaries are crisp, membership is based on common features, and relations between categories are governed by logical operations. Second, we have "fuzzy" sets, where category boundaries become vague, membership is graded, and relationships between categories become a matter of degree. Third, we have the "protoypical" model, which stresses the role of category exemplars and shifts focus from membership to degrees of fit. Finally, we have categorization in terms of "idealized cognitive models," which "motivate" the creation of categories through various forms of "chaining" and "extension."

Despite their differences in emphasis, the accounts of categorization in terms of fuzzy sets, prototypes, and idealized cognitive models all depart significantly from the classic view. They challenge basic assumptions that have underpinned both scientific and common sense theories of classification (some would say for millennia).

Once we allow that categorization proceeds in terms of prototypes, the assumption that categories express common (essential) features that distinguish entities one from another is undermined. Once we allow categorization to emerge at basic level, prior to integration into superordinate categories or refinement into subordinate categories, then the idea of categories as logical primitives becomes suspect. Nor is that idea compatible with the argument that basic level categories are grasped as wholes but with internally structured parts. Once we allow that categorization is achieved through human interaction, reflecting our bodily experience and manipulation of the world, then the assumption that categories objectively correspond to a natural order of things can no longer be sustained. If categorization is based on human experience, knowledge and imagination, including the use of imagery and metaphorical and metonymic extensions, then the relations between categories become "conventional" (though not arbitrary) and can no longer be reduced to logical formulae.

The more positive message is that while the processes of categorization may not be strictly logical, neither are they entirely arbitrary. Categorization proceeds in terms of cognitive processes that can be rendered explicit and whose adequacy can be assessed in terms of the underlying cognitive assumptions employed. A first step might be recognition of the various processes involved in categorization. Are categories based on propositional structures or informed by prototypical images? It might also involve identifying the aims of categorization (for example, prediction or inference) and making more explicit the grounds (cue or category validity, for example) on which these can be realized. In Lakoff's terms, it also involves identifying the underlying conceptual models and making explicit their metonomyic or metaphorical extensions. All of this reinforces and strengthens a central theme in grounded theory—that categories are irreducibly conceptual.

As we have seen, though, grounded theory involves a mix of messages, partly derived from the orthodox assumptions of quantitative sociology but also animated by the "naturalist" or "common sense" perspectives of the interactionist school. However, the assumptions that underpin a common sense perspective (at least in western societies) are not necessarily at variance with those of classic theory. In a common sense perspective, the world can also be divided up into "natural kinds" of objects that can be distinguished and identified in terms of their common features. The common sense view tends to identify classification with basic level categories and to assume that these have an "objective" identity.

This lack of variance is not surprising, if, as Lakoff suggests, the folk model of categorization itself "evolved into the classical theory of categorization" (1987, p.118) as (for example) in the way "scientific classification in biology grew out of folk classification (1987, p.34). The basic unit of classification in biology (the genus) was not objectively determined, Lakoff argues, but psychologically motivated to allow for ease in perception and description. Thus a biological taxonomy based on identifying similarities (that is, using "phenetic" criteria) could reinforce the common sense discriminations from which it initially sprang.

On the other hand, a taxonomy based on evolution (using "cladist" criteria) might categorize together elephants and lungfish, or categorize crocodiles with birds rather than other reptiles, suggesting that many common sense categories (such as zebras and fish) have no "objective" existence (Lakoff, 1987, p.185–195). The classic account of classification is sometimes defended by giving up on common sense altogether and taking refuge in categories constructed (or perhaps imposed) by science. However, Lakoff argues that—at least in regard to biology—science does not support the classic view. Taking Mayr's evolutionary concept of a species as an example, he shows that this departs in several ways from the classic account. For example, species are not defined in terms of intrinsic common properties, but relationally with respect to other groups, and also to time and place; while the evolution of one species into another means there may be no distinct point at which to distinguish them (Lakoff, 1987, pp.188–192).

In this clash of biological taxonomies we can find an echo of a related conflict between the role of "in vivo" and "scientific" categories in social theory. In formulating grounded theory, Glaser and Strauss likewise distinguished between two kinds of categories:

> As categories and properties emerge, the analyst will discover two kinds: those that he has constructed himself . . . and those which have been abstracted from the language of the research situation (1967, p.107).

Glaser and Strauss suggest that the analytic categories tend to provide explanations of processes and behaviors described by the in vivo categories. Later Glaser also contrasts analytic and in vivo codes (or categories) in terms of the imagery they employ. He suggests that in vivo codes convey meaning through imagery as "they have a very vivid imagery with much local interpretive meaning"—which means they have "grab" or "resonance" for people in this and similar settings. They are also familiar, easy to use, and sufficiently precise in meaning. By contrast, analytic codes are described as having "little imagery"—"some analysts think that the flatter they are the more scientific and less impressionistic" (Glaser, 1978, p.70)—but necessary to extend the scope of analysis beyond "local meanings."

Neither Glaser (1978, p.70) nor Strauss (1987, p.34) seem entirely happy with this contrast between vivid in vivo terms with "local meaning" and the "flatter" analytic constructs with wider import. They both imply a preference for analytic constructs with more imagery or resonance than conventionally allowed. Glaser even suggests a study can only take a limited dose of analytic concepts before it becomes "too rhetorical, contrived, airy or wordy" (1978, p.71). This inclination for richer or more resonant concepts is reflected in the suggestion that emergent concepts should be "both analytic and sensitizing"—designating "general properties" while at the same time being "vividly sensitizing or meaningful" to the people concerned (1967, pp.240–1). Categories that are both sensitizing and analytic, it is argued, can thereby provide a bridge between the theoretical thinking (of the sociologist) and the practical thinking (of lay people).

These considerations soften but still reflect schematic contrasts between in vivo and analytic thinking—between vivid (but local) and flatter (but wider) meanings or between analytic and practical thinking. In grounded theory, theory has to reflect but also transcend the local context. Theoretical scope and import tend to remain the province of a classic perspective in which categories are analyzed in terms of properties and relations. The sensitizing functions of categorization are confined largely to the locus of practical thinking and local application. The in vivo categories are useful for generating categories, but these must then be subjected to formal analysis. Thus Strauss and Corbin suggest that words and phrases used by informants are an important source of category names, giving as an example the description of a nurse as the "tradition bearer" of a ward:

> Now the term "tradition bearer" is a great name for a category. It's catchy, suggesting and summarizing all the things that the head nurse said about this person. In using the term we not only have a good term but we will, as with all categories, then go on to develop that category, beginning with listing some of its properties . . . which can then be dimensionalized (Strauss & Corbin, 1990, p.69).

So grounded theory retains a classic mode of categorization as an overall framework for generating theory, but modified in significant ways by the practical contexts from which theory emerges and to which it applies.

The challenges we have noted to the classic account of categorization suggest at least two problems with this approach. One problem is that it tends to separate analytic thinking from and elevate it above practical thinking, while leaving the relationship between the two unexamined. A second (and related) problem is that it presents analytic thinking in classic terms, notably the concept-indicators model, with little or no recognition of the role of fuzziness, gradation, prototypicality or cognitive idealizations in the production of knowledge.

The first problem can be illustrated by contrasting the sensitizing/analytic split in grounded theory with Lakoff's account of the relationship between basic-level (or common sense) and scientific knowledge. Lakoff argues that scientific knowledge depends (to a large degree) on the technological extension of basic-level perception. For example, microscopes (or telescopes) render the invisible visible. We "see" cells (or galaxies) previously hidden from our basic-level perception:

> Knowledge that we are confident of can grow because we see science as giving the potential to extend our basic-level perception and manipulation very much further, perhaps indefinitely further. . . It is the technological extension of basic-level perception and manipulation that makes us confident that science provides us with real knowledge (Lakoff, 1987, p.298).

Of course, our confidence in such technological extensions depends in part on the possibility of someone knowing how and why they work. Knowledge depends on such understanding, but also on the understanding that stems from our ability to perceive, experience and manipulate the products of technology in a reliable and consistent way. I may not know how the autofocus facility on my camera works—it probably depends on an application of the rules of fuzzy logic (Kosko, 1994, p.182)—but I can immediately "see" the results through the viewfinder.

In Lakoff's account, basic-level knowledge is not conceived (and circumscribed) as local and practical. It is rather the universal bedrock (at least in terms of human experience) on which scientific knowledge is based. Indeed "the existence of directly meaningful concepts . . . provides certain fixed points in the objective evaluation of situations" (Lakoff, 1987, p.300). Scientific knowledge must be "coherent with our basic level perceptions" as well as accepted by the scientific community in order to be generally accepted—in this respect, scientific knowledge is like ordinary knowledge (Lakoff, 1987, p.299). Both ordinary and scientific knowledge are not absolute but relative to our understanding and especially our basic-level understanding of experience.

Note that Lakoff is not reifying common sense here. Experience and understanding are structured as they are because the "world is as it is" but also because "we are as we are." Common sense categories are not "absolute" in the sense that they have direct correspondence with an objectively given reality; they are a product of human interaction with our environment. Human understanding is conditioned by our experience of that environment—but it remains our understanding, shaped by "our capacities for gestalt perception, mental imagery and motor movement" (Lakoff, 1987, pp.302–3). Our common sense knowledge is good enough for most purposes but it is always subject to reinterpretation and revision. For example, evolutionary theory has much enriched our knowledge (and categorization) of the relationships between species. Even such abstract knowledge as the theory of relativity can be absorbed into (and transform) common sense knowledge through practical applications, such as navigation at sea.

The sensitizing and practical bent of grounded theory points in a similar direction, but it is overlaid by an unreconstructed view of analytic categories as transcending everyday knowledge. This ignores the extent to which everyday categories are already theoretical and disguises the extent to which analytic categories are themselves experiential—that is, rooted in, derived from and ultimately adjudged meaningful in terms of basic level knowledge and conventional extensions. On the other hand, the argument that categories should be both analytic and sensitizing can be sustained more easily once we acknowledge the role of cognitive models and conventional extensions in the process of categorization.

The second problem I identified concerned the presentation of analytic categories in the classic terms of correspondence with objectively given entities, along with their distinctive properties and relations. This seems implicit in the separation of an analytic core from sensitizing description; and also in the emphasis on the discovery or emergence of categories in the research process. In grounded theory, meaningfulness in analytic terms seems to be equated with the classification of categories, properties and dimensions that directly correspond to the world as observed (or, at least, to the data accumulated through theoretical sampling). This seems to assume precisely what more recent accounts of categorization have set out to challenge: that categorization is based on identifying and distinguishing common characteristics.

IN CONCLUSION

To recap: categories may lack sharp boundaries; membership may be assigned on the basis of family resemblance rather than common features; membership may be graded, so that some are "more" members than others; and categories may acquire meaning through image-schematic, metaphorical, and metonymic extensions, rather than correspondence. The "common features" perspective fails to take account of how categories are actually assigned and used in the production of knowledge.

This need not be much of a problem, though, if the classica approach not only corresponds to our "folk theories" of classification but also yields useful results—in other words, if it has heuristic value. This raises the question of whether the challenges to classic classification offer not only (better) accounts of how we categorize but also have implications for how we think.

Applications of fuzzy logic have been most extensive in engineering where they have improved control systems in products ranging from air conditioners, refrigerators, and ovens to washing machines (not to mention elevators and subway systems). There have also been applications to schedule tasks, monitor health status, and manage investment portfolios (Kosko, 1994, pp.186–7). Most of these applications have developed in Japan and the Far East, where fuzzy logic has received a far more sympathetic response than in western science, still wedded to Aristotelian logic. The extension of fuzzy thinking and mathematics to social issues is relatively underdeveloped. Though the social sciences abound with fuzzy concepts they do not abound with methodologically innovative mathematicians (McNeill & Freiberger, 1994, pp.95–7). Nevertheless the potential of fuzzy logic for dealing with social issues such as cohabitation, retirement or unemployment (instead of insisting, for example, that people must either be "employed" or "unemployed") seems significant. Fuzzy logic also offers some prospect of reconciling the traditional tensions between qualitative and quantitative analysis, by bringing the mathematics of fuzzy sets to bear on real-life problems often created by the categorical burdens of bivalent logic.

Back (1997, p.49) suggests that fuzzy logic allows the development of new forms of logic and algebra "that effectively describes actual decision processes or linguistic meaning." However, he adds (1997, p.49) that this seems too radical a departure from classical assumptions, and so social scientists have been reluctant to abandon the use of arithmetic and the analysis of real variables.

Thinking in terms of prototypes is not an accidental nor an incidental aspect of categorization. Rosch suggests that prototypes offer the maximum information for the least cognitive effort. Like the sensitizing concepts of Glaser and Strauss, they provide a quick, memorable and incisive way of accessing or summarizing relevant information. Given that most concepts are fuzzy, prototypes provide an easy method of "knowing what we are talking about" while ignoring the hazy boundaries where trouble lies. However, information about good exemplars is more likely to be generalized to non-representative members than the reverse. For example, a study by Rips in 1975 (cited by Lakoff, 1987, p.42) showed that people more readily believe a disease spreads from robins (good examples) to ducks (not so good) than from ducks to robins. Prototypes are not innocent glosses on categorization: they act as "cognitive reference points" and form a basis for inference (Lakoff, 1987, p.45):

> It is normal for us to make inferences from typical to nontypical examples. If a typical man has hair on his head, we infer that atypical men (all other things being equal) will have hair on their heads. Moreover, a man may considered atypical by virtue of not having hair on his head. There is nothing mysterious about this. An enormous amount of our knowledge about categories of things is organized in terms of typical cases. We constantly draw inferences on the basis of that kind of knowledge. And we do it so regularly and automatically that we are rarely aware that we are doing it. Reasoning on the basis of typical cases is a major aspect of human reason (Lakoff, 1987, p.87).

This suggests that we need to take account of what we know of this and related aspects of categorization, such as factors influencing ease and speed of discrimination between alternatives. I say "what we know" because the evidence about prototype effects does not entail any particular explanation of their role in categorization and is subject to a variety of different interpretations.

Lakoff's own account of prototype effects, in terms of idealized cognitive models, may also have significant implications for categorization in grounded theory. It focuses attention on the underlying cognitive models (or assumptions) that "motivate" categorization. This to some extent may run against the grain of grounded theory, with its heuristic of "discovery" and continual caution regarding the use of "borrowed" categories. In grounded theory innocence is preserved and bias precluded by allowing categories to emerge from (and hence correspond to) the data. But Lakoff's analysis suggests that such innocence is impossible to achieve. We think in terms of categories and our categories are structured in terms of our prior experience and knowledge. If most of our categories and much of our thinking is shaped by the structuring and mapping processes that Lakoff identifies, then we would do better to reflect critically upon the underlying models (and conventions) that invest our categories (and their relationships) with meaning.

However, these contributions to the theory of categorization have all been made after the formal launch of grounded theory in 1967. At that time, the classic theory of categorization ruled the day relatively unchallenged. The explanation of categorization Glaser and Strauss offered in terms of a concept-indicator model is consistent with the tenets of classic theory. This also seems to inform claims that categories emerge from experience in ways that are unproblematic. Insofar as grounded theory is based on the classic theory of classification, it cannot ignore the challenges implied by the alternative accounts that have since developed.

In some ways, though, grounded theory already opens the door and points the way to these new issues of categorization—through its concern with sensitizing categories, and its concern for the practical origins and relevance of categories. Perhaps the problems of generating and using categories in grounded theory can therefore be readily reassessed in light of the new challenges that have emerged to the classic model.

CHAPTER 5

Coding

Coding. . . is an essential procedure. . . The excellence of the research rests in large part on the excellence of the coding.

(Strauss, 1987, p.27)

In shifting our focus in this chapter from categorization to coding, we seem to move from the magical to the mundane. With categories we impute meanings, with coding we compute them. The former involves a creative leap, for "comprehending experience via metaphor is one of the great imaginative triumphs of the human mind" (Lakoff, 1987, p.303). The latter involves reduction and ready reckoning. Ironically, the language of grounded theory, combined with growing prevalence of computing applications, has helped to promulgate and popularize coding as a key process—for some perhaps the key process—in qualitative analysis. Indeed, one suspects that the rapid rapprochement of computing techniques with grounded theory methodology owes something to its explicit avowal of coding as a qualitative procedure.

In this chapter, we consider the way coding is conceived in grounded theory. This will take us through the main paths of the methodology—through open, axiel and selective coding to the selection of a core category around which to organize the analysis. It will take us into the controversy between Glaser and Strauss over the use of the coding paradigm, though we return to this issue when we consider the analysis of process in a later chapter. Meantime I consider Glaser's distinction between substantive and theoretical codes and his amplification of various coding families. Although the distinctions between different types or stages of coding in grounded theory have become axiomatic, I suggest that the analytic process they express can easily be misconceived. In particular, I emphasize the need to retain a holistic sensibility, expressed in the idea of generating category "strings" rather than individual codes that then need to be reconnected. This emphasis is consistent with our earlier discussion of categorization, notably Lakoff's explication of meaning in terms of idealized cognitive models. Although categories are also presented in grounded theory as irreducibly conceptual, ambiguities arise in its account of coding and the extent to which it is dictated by what is in the data.

CODING IN GROUNDED THEORY

In their original text, Glaser and Strauss aimed to establish (or at least improve) the legitimacy of qualitative methods by borrowing from the language and procedures of quantitative research. The term "coding" is usually applied to procedures for managing responses to "precoded" questions in survey research. In this context, the conceptual work (involving the identification of relevant response categories) precedes the mechanical task of assigning codes to responses to facilitate computational analysis. So coding in this context refers only to the mainly (but even here not invariably) mechanical tasks of identifying and assigning the appropriate codes to responses. However, at least the conceptual work is already done. In qualitative analysis, the task of conceptualizing responses remains to be accomplished. Yet the implication of coding data may be that somehow the data can be coded even before it has been conceptualized or analyzed.

To the contrary, Glaser and Strauss vigorously assert that coding and analysis proceed jointly in grounded theory—indeed, Strauss later defines coding simply as "the process of analysis" (Strauss & Corbin, 1990, p.61). In this respect, they contrast grounded theory with several other methods where coding and analysis are not combined. Most emphatically, they reject the method of coding data "into crudely quantifiable form" in order to test hypotheses, since they are interested in generating theory rather than verifying it. Unless the researcher's aim is to establish "proofs," incidents need only be coded when they point to a "new aspect" of a category (Glaser & Strauss, 1967, p.111). In other words, coding is governed only by theoretical relevance and it is not concerned with the accumulation of supporting evidence. An objection from Merton that the accumulation of evidence and its analysis (for example, through frequencies and cross-tabulations) can itself be useful in generating theory is noted (in a footnote) but otherwise ignored. This is a point to which I return.

Glaser and Strauss also reject as unsystematic the method of generating theory without coding data at all, but by merely "inspecting" it. By contrast, they advocate coding as a method of making conceptualization explicit.

> The purpose of the constant comparative method of joint coding and analysis is to generate theory more systematically . . . by using explicit coding and analytic procedures (Glaser & Strauss, 1967, p.102).

Note that "systematically" refers here only to the explicit use of concepts and not the extent to which coding is done. This half-way house between full coding and no coding at all could be called "partial" coding. With partial coding, there is no attempt to provide an exhaustive analysis of evidence with the aim of refuting (or supporting) hypotheses:

> . . . the constant comparative method cannot be used for both provisional testing and discovering theory: in theoretical sampling, the data collected are not extensive enough, and, because of theoretical saturation, are not coded extensively enough to yield provisional tests. . . They are coded only enough to generate, hence to suggest, theory (Glaser & Strauss, 1987, p.103).

The key function of coding, then, is to generate rather than to test theory. How coding can contribute to the generation of theory is elaborated in the later texts. Let us look first at how coding contributes to this task in grounded theory.

OPEN, AXIEL, AND SELECTIVE CODING

In the version presented by Strauss and Corbin, grounded theory confronts the problem of developing analytic focus by emphasizing particular phases of coding. These phases are divided into open, axiel, and selective coding. As the theoretical focus sharpens, analysis passes from one phase to another.

Open coding is defined as "the process of breaking down, examining, comparing, conceptualizing and categorizing data" (Strauss & Corbin, 1990, p.61). In a further clarification, it is defined as "the part of analysis that pertains specifically to the naming and categorizing of phenomena through close examination of data" (Strauss & Corbin, 1990, p.62). This is presented as "the first basic analytical step" from which everything else follows. Everything else includes the phases of axiel and selective coding. Axiel coding is defined as "a set of procedures whereby data are put back together in new ways after open coding, by making connections between categories" (Strauss & Corbin, 1990, p.96). Selective coding is defined as

> . . . the process of selecting the core category, systematically relating it to other categories, validating those relationships, and filling in categories that need further refinement and development (Strauss & Corbin, 1990, p.116).

The core category in this context is described as "the central phenomenon around which all other categories are integrated" (Strauss & Corbin, 1990, p.116).

These various phases of analysis can be summarized as:

1. Categorizing the data (open coding)
2. Connecting categories (theoretical or axiel coding)
3. Focusing on a core category (selective coding)

This may be regarded as representing the common folklore of qualitative analysis. First construct categories, then connect them, and finally organize them around an integrating theme. Or first analyze, then synthesise, and finally prioritize. How else could it be done?

One of the conventional features of this account is the separation of the processes of categorizing data and connecting categories into distinct phases. In open coding, we are told, "the data are broken down into discrete parts" and then "closely examined" and "compared for similarities and differences" (Strauss & Corbin, 1990, p.62).

This process is informed by asking questions of the data, though Strauss and Corbin part ways with Glaser on what these questions might be. According to Strauss and Corbin (1990, p.77) "the basic questions are Who? When? Where? What? How? How much? and Why?"—some of the stock questions of social science. Glaser (1992, p.51), on the other hand, suggests a different three:

1. What is this data a study of?
2. What category or property does the incident indicate?
3. What is the basic process that "processes the main problem that makes life viable in the action scene"?

In the last question Glaser seems to be asking about the importance of the data—that is, whether (or how) it sheds light on the "basic social processes" under study. Glaser's questions are at a much higher level of generality than those suggested by Strauss and Corbin, which, Glaser suggests, "force" the data in particular directions rather than allow categories to "emerge" from what is there. We return to this issue later.

Meantime, let us return to the separation of categorizing data from connecting categories (which is another issue on which Glaser expresses his disapproval of what Strauss and Corbin propose). The proposed fragmentation of data into discrete parts may seem unexceptional—since, after all, this is what software programs for qualitative data analysis commonly support. Parts of the data that seem relevant to the analysis are earmarked for further examination. The rest of the data are not (necessarily) forgotten (though some parts may be discarded as utterly irrelevant) but generally becomes the backdrop against which the earmarked segments are viewed. The data segments are categorized, both in terms of their relevance to the analysis (how they contribute to answering those questions) and through comparison with other segments and with the emerging categories.

Although Strauss and Corbin (1990, pp.72–3) suggest that open coding can be carried through at different levels from a line-by-line analysis, through sentence or paragraph coding, to coding an entire document, the emphasis nevertheless is on "close examination" through a line-by-line analysis—a fine-grained analysis that may even focus on particular words or phrases. This is seen as "the most detailed type of analysis, but the most generative." Coding by sentence, paragraph or document is presented as a way into or of following up (but not as an alternative to) this more detailed analysis. It is, of course, impossible to sustain such a fine-grained approach to the whole of any set of data of even moderate proportions—such as a grounded theory approach is bound to produce. It is appropriate, therefore, only at the outset of the analysis, when the initial categories have still to emerge.

Holistic Understanding

This approach may seem to forfeit the potential of a more "holistic" view of the data. In relation to coding, the occasional benefits of a broader view are recognized, but more as an aid to a fine-grained analysis than as worthwhile in its own right. The data must be "broken down" for close examination. This, of course, is entirely in keeping with an analytic approach, which involves "a process of resolving data into its constituent components, to reveal its characteristic elements and structure" (Dey, 1993, p.30). The word "analysis" derives from the prefix "ana-" together with a Greek root "lysis" meaning "to break up or dissolve" (Bohm, 1983, pp.125 and 156). In an earlier text, I suggested that "without analysis, we would have to rely entirely on impressions and intuitions about the data as a whole" without the benefit of "the more rigorous and logical procedures" which analysis can offer (Dey, 1993, p.30).

Now I am no longer so sure. The body once dissected cannot be resurrected. Something may be gained through analysis, but also something may be lost. And that loss may be too readily diminished or dismissed, if it is seen (as in my earlier text) as the product of mere "impressions and intuitions." But what can a holistic view offer? And how can it be sustained?

> Analysis: *Resolution into simpler elements; after due consideration (as "in the final analysis").*
> Holism: *Tendency in nature to form wholes that are more than the sum of the parts by ordered grouping.*

The first part is easier to answer than the second, since it involves the classic contrast between the wood and the trees. But this is not just a question of "seeing" an object, as in the aphorism "not seeing the wood for the trees." This after all is simply a function of scale. The point is rather that the whole wood may display complex processes (a "life" of its own) that cannot be explained in terms of the individual trees. For example, the ecology of a forest is not the sum of the ecology of each individual tree. The trees in a forest behave differently from trees in isolation—for example, in their competition for light. The forest may have its own cycles of reproduction and regeneration. You can manage and "harvest" a forest in a way you cannot with individual trees. And you can take a walk in a wood or get lost in a forest (and always do in fairy stories) but never (except in fairy stories—"Jack and the Beanstalk" springs to mind) in an individual tree. Thus there are processes that we can only understand if we recognize the forest as a forest and refuse to analyze it in terms of individual trees.

Similarly, our bodies (and we ourselves) are more than the sum of individual parts: "you can study arms and legs and organs and other parts and still not know how a human behaves" (Kosko, 1994, p.108). The groups, institutions, cultures, economies, and societies in which we live also display complex behaviors that cannot be understood in terms of individual action. Like the trees growing in the forest, the individuals themselves are already shaped by the social environment in which they act. There can be no performers without an audience, no banks without customers, no producers without consumers, no writers without readers (we hope!)—and so on. And in between the performer and the audience, we have the impresarios, the stage managers, stage and prop designers, curtain, light and sound operators, the reviewers, and a host of other supporting actors all contributing their bit to the complex process of producing a show. Put all these people together (fulfilling these roles) and we do not just get a bunch of people together—we get (we hope) a performance.

Kosko suggests that the whole is more than the sum of its parts because "system complexity exceeds subsystem complexity" (1994, p.108). But in another sense, the whole is also simpler than each of the parts. It is easier to appreciate the performance as a whole, than to comprehend the interrelated sequence of complicated tasks and roles required to produce it. I experience the performance from my position in the audience. In the same way, we can grasp the nature and function of our furniture (chairs, lights, desks) without understanding the intricacies of its construction—what a dovetail joint is or how the wiring circuits work.

Experience and Prior Knowledge

As I noted earlier, Lakoff suggests that this knowledge is not an abstraction constructed through some complicated logical process of induction from particulars—it is understood directly in the light of our bodily experience. When I go shopping, I have a direct understanding of what a shop is based on my bodily experience of entering the shop, filling a basket with goods, and (the difficult bit) of parting with cash in exchange for my purchases. Of course nowadays I do not actually part with cash at all—I give them a small card that they swipe through a machine and suddenly my bank account is debited the cost of my acquisitions.

How can I understand this transaction? Fortunately I do not need to grasp the intricacies of the telephone system that communicates between shop and bank or the complexities of the cash economy within which this transaction occurs. I do not even have to understand how my finances may be affected in the short or longer term by the purchase of items by credit card (some people never do). So what do I need to know? I have to remember to take the card with me when I go shopping. I have to know that I must give it to the cashier—though (s)he will soon tell me if I forget. That is a minimum. I probably have to know, too, that there is a direct connection between the debit that will appear in my bank statement I get at the start of each month and the transaction that has just occurred. Otherwise I may think the way these cards can produce money at cash dispensers is little short of magic (as one of my children believed). All this knowledge is based on my (bodily) experiences of taking the card, passing it over, watching it being swiped through themachine, and receiving and reading the bank statement.

Of course, my knowledge of card transactions may also have other sources. Even before cards became common, I had read about their experimental introduction in a small region of France under another headline that promised the demise of the cash economy. I also obtained a rather thick wadge of documents on first obtaining a card, and I have a dim recollection of wading (or, rather, skimming) through some pages of very small print. This information gave me some idea of what was involved—though I doubt if it was essential, as my children have happily acquired and used cards (through emulation) without this benefit. And, as everybody starting work soon realizes, there is a world of difference between the abstract knowledge in books and the practical knowledge required for and acquired in everyday experience—between reading what to do, seeing others do it, and doing it for yourself. I hasten to add that I am not intending to denigrate abstract knowledge or books here—or I would not be writing one. My aim is rather to suggest that much of our holistic knowledge of our social world may be based on direct understandings that are acquired through our bodily experiences in that world.

IMPRESSIONS, INTUITIONS, AND DIRECT UNDERSTANDINGS

If this is so, then it suggests that our holistic knowledge has a firmer basis than "mere" impression and intuition. These we tend to think of as ephemeral or peripheral.

Impressions are not the stuff of direct experience but only the superficial imprint of what has passed—and so they can be easily misinterpreted. Unlike direct understanding, impressions are liable to be vague and misleading. We use the term in those circumstances where appearances may have led us astray. Impressions must therefore be checked against further experience.

> Impression: *1. Impressing (of mark) 2. Mark impressed. 3. Effect produced (especially on mind or feelings); representation of this by artist, mimic etc. 4. Notion, (vague or mistaken) belief, impressed on the mind (as in "I was under the impression")...*
> Intuition: *Immediate apprehension by the mind without reasoning; immediate apprehension by sense; immediate insight.*

Intuition seems more endearing, since it promises to deliver a deeper truth without the tedious hassles of logical deduction. Here truth is divined immediately but not without the gift of sudden inspiration and insight. The trouble is that such gifts are bestowed sparingly and cannot be relied upon for routine, mundane everyday, practical requirements. Sadly, intuition cannot be switched on and off like a light whenever you would like a little illumination.

Unlike impression and intuition, direct understandings that are acquired through our bodily experience in the world may be both rich and routine. The routine comes with the everyday nature of our understanding, which makes our remarkable achievements (such as shopping) unremarkable. We acquire this kind of understanding not through revelation or instruction but through everyday activity. It has always struck me as bizarre how often we are represented as living in our heads, rather than living in the world as experienced. I blame Descartes—"I think therefore I am"—for turning our world upside down. Descartes could only think that thought because there was a Descartes to think it—who lived, ate, slept, and (when time permitted) did a bit of philosophizing. Descartes (typical male!) apparently forgot the minor practical details (being conceived, being born, being brought up, acquiring a language, etc.) that allowed him to think his thought. But I am sure his parents did not forget—and we need not either.

The richness comes with the depth of our understanding, which is dug deep through our interactions with the world as inherited. We do not make the world up entirely to suit ourselves (or I guess I would not be writing this book and you might not be reading it). The world we inhabit is the world we inherit. We are products of evolution—not, perhaps, such prize specimens as we fondly imagine, but the result nonetheless of an evolutionary process spanning not millennia but millions of years—some five million or so in the case of human beings. Some things change slowly, even over such a long timescale. We still share some 98% of our genetic inheritance with the common chimp of Africa and the pygmy chimp of Zaire. In some respects, therefore, we are best conceived of as "the third chimpanzee" (Diamond, 1992). But in other respects, things have changed dramatically because "the particular trick of human beings is not to forget but to accumulate, over time, all the tricks of other people and of previous generations" (Tudge, 1996, p.2). These "tricks" include language, technology, trade, agriculture, and industry; and as these tricks accumulate, they shape the cultural and social environment in which we find ourselves. My shopping, therefore, mundane as it may seem, is also a product of this immense process of evolution: it is rich as well as routine.

A holistic view of our data may therefore be informed by routine but rich understandings of the world that we inhabit and inherit. If so, this is likely to prove neither ephemeral nor peripheral to our account of the data. Indeed, the very possibility of producing an account (which will be understood) may depend on such understandings, that involve grasping activities, processes, or events as complex wholes rather than breaking them down into constituent parts.

At this point, it may be useful to recall the importance of practical knowledge of the social world under investigation in how that knowledge is generated and confirmed. Underpinning and guiding the process of coding is a wealth of knowledge gained, as Glaser and Strauss put it, from "getting by" in the social setting. Perhaps in grounded theory this background knowledge provides a more holistic frame of reference within which the more analytic coding proceeds?

Category Strings

So far we have considered the procedure of breaking down or fragmenting data into different bits or segments. I have suggested that something may be gained from this process (we can see the data segments in a new context, based on comparison) but also that something may be lost (in terms of a more holistic understanding). This procedure itself stems from the injunction to categorize first (through open coding) and then to connect categories later (through axiel coding). This raises a further puzzle. For how can we create or assign categories without already identifying and taking account of the connections between them?

> Category: *Class, division; one of a possibly exhaustive set of classes among which all things might be distributed; one of the a priori conceptions applied by the mind to sense-impressions; any fundamental philosophical concept.*

A categorization, after all, is not an isolated act. It is but one moment in a process of conceptualizing the data. The construction of a category or the appropriateness of assigning it to some part of the data will undoubtedly reflect our wider comprehensions—both of the data and what we are trying to do with it. The researcher (who brings to categorization an evolving set of assumptions, biases, and sensitivities) cannot be eliminated from this process.

Glaser (1992, p.47) nicely illustrates this point in making his (retrospective) criticism or the use of "pet" codes—suggesting that Strauss tended to find "pacing in everything." Given a pet code, he comments that "an analyst can be quite skilled at using the code on any data irrespective of what his data is a study of" (Glaser, 1992, p.47). Thus Strauss is accused of using categories that reflect his own preoccupations. But could (or should) it be otherwise? Glaser's comment perhaps confuses evidence with inquiry—data cannot be a "study of" anything in themselves, for a study is something we undertake (not the data). Even if we accept the (doubtful) proposition that categories are discovered, what we discover will depend in some degree on what we are looking for—just as Columbus could hardly have "discovered" America if he had not been looking for the "Indies" in the first place.

In any case, categories cannot be considered in isolation. Categories acquire their meaning in part from their place in the wider scheme of things, just as my cash card cannot be conceived apart from banks, shops, telephones, and even the postal system. Rosch discusses this issue in relation to identifying attributes. She points out that some attributes such as the seat of a chair can only be understood in terms our knowledge of the overall object. She also notes that discriminations among objects may depend on their place in a larger taxonomy. Thus describing a piano as "large" is meaningful in relation to a superordinate category—furniture; it may be "small" in relation to some other objects, such as buildings. Other attributes seem to require knowledge of our own activities, such as "eating" at a "table." On the basis of her research, Rosch concludes that:

> . . . the analysis of objects into attributes was a rather sophisticated activity that our subjects (and indeed a system of cultural knowledge) might well be considered to be able to impose only after the development of the category system (Rosch, 1978, p.42 her emphasis).

If Lakoff is right, then the categories we use will reflect the underlying cognitive models that "motivate" their meaning. When we categorize data, we do not construct an isolated number of categories that have no relationship to each other. We have to use categories in conjunction, so that we can be sure of capturing much of the significance of the data. We have to consider how our categories overlap or complement each other so that we can eliminate needless redundancies (such as using categories with different names for the same purpose). Each category has to be considered in relation to the others so that we can ensure consistency in the way we conceptualize. We have to think not only about how the categories fit the data, but whether they do so in a way that suits our wider conceptual aims.

Therefore we think in terms of category sets rather than single categories. Although I used the term "category set" in an earlier text (Dey, 1993, p.97) to emphasize this plurality, I now think this term is not entirely appropriate, since categories may still be conceived as isolated members of the set, whose only common feature is their use in the particular study. It might be more helpful, therefore, to think in terms of category "strings," with the individual categories conceived as particular points (or knots) on the string (Fig 5.1). This metaphor conveys more emphatically the way categories can help us to conceptualize the data through their connections as much as through their individual assignation. In constructing categories, we also create the string that connects them and through which they help us to capture the data. Of course, strings can also be folded, looped, and otherwise manipulated in order to create the conceptual "webs" or "nets" through which to develop our analysis.

Therefore, in grounded theory, the division between open coding and axiel coding needs to be treated with caution. We should not take this as an exhortation to categorize first through open coding and then consider connections between categories later, in axiel coding. Of course, if open coding "fractures" the data into categories (and their properties and dimensions) axiel coding is needed to put the data back together again (though in new ways), for example, "by making connections between a category and its subcategories" (Strauss & Corbin, 1990, p.97). But if open coding already requires attention to category strings, then this can be no more than a shift of emphasis from particular points in the string to the links between them.

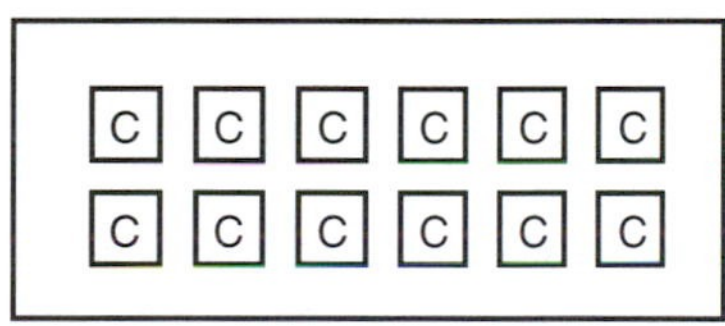

Categories in a set

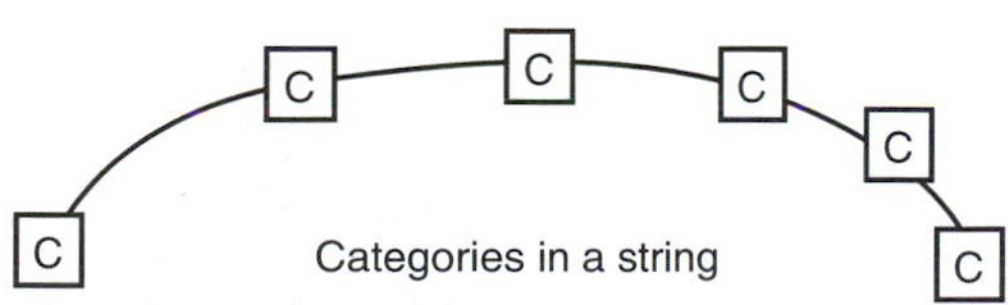

FIGURE 5.1 Categories in a set or as points in a string.

Axiel Coding

The way categories are to be connected in axiel coding is set out by Strauss and Corbin in terms of a particular coding paradigm. Strauss suggests this paradigm "is central to coding procedures" (1987, p.27). It is not enough to "code an event qua indicator as an instance of a category" as the researcher also "needs to code the associated subcategories." Here the term "sub-categories" is used (confusingly I think) to refer to the "conditions, context, action/interactional strategies and consequences" of an event. Thus when Strauss and Corbin refer to "making connections between a category and its subcategories" they mean analyzing an event in terms of these conditions, context, strategies and consequences.

The coding paradigm which Strauss (1987, p.27) suggests "quite literally becomes part and parcel of the analyst's thought processes" provides a reminder "to code data for relevance to whatever phenomena are referenced by a given category" as follows:

- Conditions
- Interaction among the actors
- Strategies and tactics
- Consequences

Strauss asserts that without inclusion of the paradigm items "coding is not coding" (1987, p.28). In other words, categories make sense only within a frame of reference which focuses on the items identified by the coding paradigm.

This coding paradigm therefore requires analysis of "conditions, context, action/interactional strategies, and consequences" (Strauss & Corbin, 1990, p.96):

> In axiel coding our focus is on specifying a category (phenomenon) in terms of the conditions that give rise to it; the context (its specific set of properties) in which it is embedded; the action/interactional strategies by which it is handled, managed, carried out; and the consequences of those strategies (Strauss & Corbin, 1990, p.97).

Most of this seems familiar enough, though one notes the absence here of any explicit reference to "causes." However, it is clear that Strauss and Corbin include the idea of "causal conditions" within the general reference to conditions.

I return to the coding paradigm when we look in the next chapter at how process is analyzed in grounded theory. Meantime let us consider the paradigm from the point of view of coding data. The paradigm requires us to connect categories in terms of conditions, context, strategies, and consequences. This may often make good heuristic sense. Such a causal perspective may well illuminate the process of interaction. But why should this paradigm be privileged? And is the privileged status of this paradigm not paradoxical given the emphasis on "emergence" and "discovery" in grounded theory? These are the questions that exercised Glaser in his critique of the coding paradigm.

Theoretical codes

One of Glaser's criticisms of the coding paradigm is that it ignored (Glaser complained) his own earlier work on theoretical coding. Instead of "forcing" the data to fit a pregiven paradigm, Glaser suggests we consider a range of theoretical options of which the proposed paradigm is only one. Glaser argues that analysts must be sensitive to "the myriad of implicit integrative possibilities in the data" (Glaser, 1978, p.73) and accordingly sets out a range of "families" of theoretical codes that can be brought into play [see inset box]. These families vary considerably in level of abstraction and also overlap—they are not meant to be mutually exclusive. The "six Cs" family is obviously kin to the proposed coding paradigm, but here figures as only one of a number of theoretical perspectives.

Glaser's Coding Families

Family	Examples
Six Cs:	Causes, contexts, contingencies, consequences, covariances, and conditions
Process	Stages, phases, progressions, etc.
Degree	Limit, range, intensity, etc.
Dimension	Elements, divisions, properties, etc.
Type	Type, form, kinds, styles, classes, etc.
Strategy	Strategies, tactics, mechanisms, etc.
Interactive	Mutual effects, reciprocity, mutual trajectory, etc.
Identity-Self	Self-image, self-concept, self-worth, etc.
Cutting Point	Boundary, critical juncture, turning point, etc.
Means-Goals	End, purpose, goal, etc.
Cultural	Norms, values, beliefs, etc.
Consensus	Clusters, agreements, contracts, etc.
Mainline	Social control, recruitment, socialization, etc.
Theoretical	Parsimony, scope, integration, etc.
Ordering or Elaboration	Structural, temporal, conceptual
Unit	Collective, group, nation, etc.
Reading	Concepts, problems & hypotheses
Models	Linear, spatial, etc.

Source: Adapted from Glaser (1978, p.81).

These theoretical codes present a wider range of perspectives on data than the coding paradigm proposed by Strauss and Corbin, though one could perhaps assimilate them with that paradigm. Although the theoretical codes are obviously preconceived, Glaser insists that they must nevertheless in some sense emerge from the data. This means that a "coding family" should be used in analysis only once indicated as appropriate by the data.

Theoretical codes often remain implicit in the way substantive codes are interrelated, and this may be a problem if it obscures the nature of the relationships being hypothesized. Awareness of the range of coding families available therefore helps to avoid the imposition of a preferred coding family (or a "pet" family as Glaser puts it) that does not really fit the data. Theoretical coding also helps "prevent the analyst from dropping and bogging down in data" (Glaser, 1978, p.73) by keeping their focus on the task of generating theory. According to Glaser:

> Theoretical codes conceptualize how the substantive codes may relate to each other as hypotheses to be integrated into a theory. They, like substantive codes, are emergent; they weave the fractured story back together again. . . .Theoretical codes give integrative scope, broad pictures and a new perspective (Glaser, 1978, p.72).

The use of a reservoir of wide-ranging theoretical codes promises to open up inquiry, bringing new questions to bear and placing the value of old questions in doubt. Nevertheless, there are some problems with this elaboration of theoretical coding, apart from the apparent confusion of very different items in Glaser's list—some of which, such as the theoretical or reading families, it is difficult to imagine as codes at all.

Issues in Theoretical Coding

First, the distinction between substantive and theoretical coding is not very clear. Glaser presents theoretical coding as "implicit" in substantive coding, suggesting that in doing the latter, one is inevitably engaged in the former. He presents theoretical coding itself as a separate activity—that of relating the substantive categories. One question this raises is whether categories at some level can be identified which do not already involve some theoretical elements: for example, such as causation, process, degree and so on. Do categories "stand by themselves" or are they not always part of a broader conceptualization that already implies relationships among the categories? My earlier discussion of category strings suggests the latter.

Second, the idea of "emergence" becomes rather clouded by the identification of a range of preconceived concepts that can relate categories. Glaser argues that theoretical codes must emerge from the data but this presumably means something different from the emergence of substantive categories (which cannot be formulated in advance). Apparently emergence of theoretical codes involves the selection of appropriate codes (from among those available) given cues in the data. It seems that this selection may at times be rather arbitrary, since more than one coding family may fit the same data.

A further problem is that the nature of the empirical cues guiding selection is not spelled out. Do these cues arise because relationships can be observed directly in the data? For example, can we observe directly relationships such as cause and effect or differences of degree? Or are these inferences, that are based on observation, but made within the context of a particular theoretical perspective? Glaser's insistence on emergence implies the former; but his talk of cues and selection suggests the latter. In either case, it is not clear whether Glaser is implying that observations are somehow pretheoretical. For example, do processes divide naturally into stages, or is this rather a construct used by the analyst to order events?

Some Questions about Theoretical Coding

- *Is theoretical coding an aspect of substantive coding" or a separate activity?*
- *How do we select among theoretical codes that all fit the data?*
- *Are the empirical cues guiding selection based only on observation?*

While Glaser's discussion of theoretical codes may sensitize us to different modes of theorizing, we also need to think more explicitly about how to make connections between concepts. For example, Sylvan and Glassner (1985) distinguish three different methods of relating properties, through sets, logic, and grammars. These methods are not unrelated, as each relates properties by using rules. But the rules do differ. The rules in sets use unions and intersections, for example, while the rules in grammars depend instead on "ordered associations" (Sylvan & Glassner, 1985, p.113). In this respect, the different types of codes which Glaser identifies may serve as a useful starting point rather than as a conclusion. As we see in the next chapter, even identifying something so apparently simple and central as a causal connection can give us plenty of problems. We therefore need to make explicit (as far as possible) the rules or at least the criteria governing the identification of properties and their relations.

CORE CATEGORIES

Another direction that Glaser takes (and Strauss follows) in developing grounded theory involves the identification of a "core category" as a key part of the process. Indeed, he asserts that "the generation of theory occurs around a core category" (Glaser, 1978, p.93). The primacy of a core category is justified in terms of both ends and means. In terms of ends, the goal of grounded theory is "to generate a theory that accounts for a pattern of behavior which is relevant and problematic for those involved"—and not what Glaser describes (perhaps rather derisively) as "voluminous description" or "clever verification" (Glaser, 1978, p.93). The aim of producing a grounded theory that is relevant and workable to practitioners requires conceptual delimitation. The focus on a core category provides a means of achieving this by ensuring "relevancy and workability" (Glaser, 1978, p.93). Categories that are not related to the core category can be excluded from the theory; other categories that are related to the core category are examined in more depth. Also, the core category conceptualizes the "basic social process" or basic problem addressed by the theory, providing an integrative framework around which the analysis develops. The use of a core category in this way is intended both to maximize parsimony and scope while ensuring that the emerging theory is "dense and saturated" (Glaser, 1978, p.93).

Strauss and Corbin also discuss the use of core categories in their account of the third phase of analysis: selective coding. This is defined as follows:

> Selective coding: The process of selecting the core category, systematically relating it to other categories, validating those relationships, and filling in categories in that need further refinement and development (Strauss & Corbin, 1990, p.116).

It is essential to choose a core category, they argue, "in order to achieve the tight integration and dense development of categories required of a grounded theory" (Strauss & Corbin, 1990, p.121). In their account, the identification of a core category follows from developing a descriptive narrative of "the central phenomenon of the study"—that is, "explicating the story line" (Strauss & Corbin, 1990, pp.116–7). This can be done by writing a brief abstract of the study—a few sentences at most—to develop a general descriptive overview of the research. On that basis, a core category which conveys this analytically can be created (by giving the central phenomenon a name) or selected from the category list "if one of them is abstract enough to encompass all that has been described in the story" (Strauss & Corbin, 1990, p.120).

As with Glaser's account, the question arises whether core categories are indicated by the data or selected by the theorist? As core categories are "discovered," it seems they are indicated by the data. As more than one can be discovered, and as the researcher has to decide which to privilege and which to demote, it seems they are selected by the researcher. In their discussion of how to choose between two competing categories, Strauss and Corbin imply that the decision to privilege of one category over another is quite arbitrary:

> The way to handle this problem is to choose one phenomenon, relate the other category to it as a subsidiary category, then write it as a single theory. Then, in another paper or monograph you can take up the second idea and do the same (Strauss & Corbin, 1990, pp.121–2).

Of course the researcher can only choose among core categories indicated by the data. But is the core category really indicated by the data, or is its identification simply the result of a heuristic requirement to delimit the emerging theory?

According to Glaser, a core category "accounts for most of the variation in a pattern of behavior" (Glaser, 1978, p.93) and therefore it has to earn its privileged position. On the other hand:

> Many studies yield two or (sometimes) three core variables. To try to write about them all at once with no relative emphasis is to denude each of its powerful theoretical functions (Glaser, 1978, p.94).

The process of identifying a core category therefore emphasizes the judgment role of the theorist. Glaser warns against premature selection, which can produce "an undeveloped, undense theory with little explanatory power" because the selected core category fails to integrate other categories effectively.

To be effective, a core category must be central, stable, complex, integrative, incisive, powerful, and highly variable (to adapt and summarise Glaser's account a little).

- It is *central* if it is related to many other categories and accounts for a large proportion of variation in the data.
- It is *stable* if it can be seen as a recurrent pattern in the data.
- It is sufficiently complex if it takes more time to saturate (identify its properties) than other categories.
- It is *incisive* if it has clear implications for more formal theory.
- It is *powerful* if its explanatory power helps the analyst to "carry through" the analysis to a successful conclusion.
- It is highly *variable* if it is sensitive to variations in conditions in terms of degree, dimension, and type.

None of this explains why only one category (or perhaps two or three at most) should meet these criteria. Why not half a dozen—or more? There seems to be no reason, a priori, why any one category should account for a large portion of the variation in the data. This is one area where the normal injunction of grounded theory—never to look for a fit in the data—seems to be suspended. Indeed, the theorist is encouraged to look for a core category from the start, though to be fair, also cautioned not to rush to judgment and to exercise patience in a search that "tests the analyst's skills and abilities" (Glaser, 1978, p.95). It involves "taking a chance" because the selection cannot be justified in advance but must be "finally proven by sorting data into a theory that works" (Glaser, 1978, p.95). This is certainly contrary to the usual cannon of grounding analysis in the data—which hardly allows for taking a gamble (even an informed one) on a theoretical hunch!

Taking one core category as a fulcrum for theory may also mislead if it excludes or underestimates the role of other important factors. The research may result in a single product rather than offer a menu of possibilities. Selecting a core category seems to involve the elimination of alternative accounts—for these are relegated to future reports. It suggests that there is no place for conflicting and contradictory explanations, which may be more or less supported by the available evidence. The logic of the analysis permits only one positive conclusion: that the core variable has emerged with some explanatory power. But it does not seem to permit that explanatory power to be pitted against the potential of alternative explanations.

Finally, there seems to be an apparent paradox in the criteria presented by Glaser for selecting a core category. On the one hand, it is presented as a powerful independent variable that accounts for much variation in the data. On the other, it is presented as a highly dependent variable that is much affected by its relationship to other factors. Usually we think of variables as either independent or dependent, and research designs are devised to examine either the factors that determine a dependent variable or the consequences that flow from an independent variable. Here the core variable is both. For example, it may be highly dependent in relation to conditions and strongly independent in relation to consequences. A focus on explaining variations in the core category, as well as variation explained by it, may give a powerful impetus to theory which thereby not only analyses relationships between categories but also takes account of the conditions under which these hypotheses may hold. But it imposes a more demanding design than research that focuses mainly on understanding either conditions or consequences.

IN CONCLUSION

For many, the distinction between open, axiel, and selective coding is a hallmark of grounded theory. The phase of open coding in particular attracts the attention and support of those who approve of the creative impulse expressed in this initial commitment to a wide-ranging and unstructured response to data. The term "open" expresses well the key role in grounded theory of a creative response to data. The axiel and selective phases of coding seem less exciting and more controversial, as they involve the introduction of a particular coding paradigm and the imposition of a more focused and structured discipline on the coding process.

The concerns I have raised do question the general analytic shift from a more open to a more focused form of coding data. This shift seems eminently plausible, particularly given the lack of any prior theoretical commitments on the part of the theorist. However, there are issues to consider with respect to each of these phases in the analytic process. With regard to open coding, I suggested that a more holistic approach might be worth pursuing and noted that this may be more firmly based on experiential and practical forms of knowledge ("direct understandings") than my own scepticism regarding impressions and intuitions had allowed. To encourage this approach, I suggested we think in terms of category strings rather than categories as concepts that "stand alone." With regard to axiel coding, Glaser's criticism of the coding paradigm as "forcing" the data raises a more general question about the role of the theorist in the construction of codes—a question that applies with equal force to Glaser's own families of theoretical codes. One question here concerned the extent to which selection of a particular coding family is an arbitrary process, if more than one family can fit the data. A similar issue arose with regard to the selection of a core category, since here too there is some ambivalence over how far this is imposed by or on the data.

One of the rationales for the development of a core category is the need to bring coding to some sort of conclusion. The focus on a core category assists in reaching a resolution to inquiry by setting some bounds to the analytic process. Another important way in which coding is bounded can be found in the concept of theoretical saturation. This raises the general question of how far coding should be taken in grounded theory, and when to stop. This vexed question is the subject of the next chapter.

CHAPTER 6

Concluding

In this chapter, I consider the problem of knowing when to stop. This is an important consideration, and one which in everyday life we often neglect to our cost. Whether driving or dating, the temptation to go "round the next corner" is well nigh irresistible. It is much to the credit of Glaser and Strauss that they pose the question not just of how to start coding but also of how to end it. The answer in grounded theory is couched (rather misleadingly, I suggest) in terms of "theoretical saturation," so this provides an appropriate point at which to start. The tension between partial and full coding of data is conveyed but not entirely resolved by this ambiguous term; and so it is not surprising that it reappears (in new guises) in the later literature on coding.

I consider several efforts to wrestle with the question of how much coding is required of data, either in grounded theory or more generally within a qualitative analysis. The question has been posed more sharply by the capacity of the computer (if not the analyst) to handle large volumes of data. The oppositions I discuss between factual and referential coding, between factual and heuristic coding, and between data reduction and complication form the framework of this on-going struggle. However, I suggest that these attempts to excise a mechanistic approach to coding are ultimately futile, as the term itself reflects an underlying cognitive model which is thoroughly aconceptual.

More positively, I also suggest that the game has already been won in the sense that no matter how mundane and elementary the tasks to which coding refers, they are inherently and inescapably conceptual in character. The point is not, therefore, to distance analysis from these tasks so much as to recognize and make explicit their conceptual character.

This perspective reopens the question of when to stop, but from a rather different vantage point. It is now possible to see the counterposition of creative and mechanical tasks as misleading and to suggest that analysis can proceed creatively through the mundane as well as the magical. This brings us to "counting" and the role that numbers can play in the analysis of meaning. I suggest that the prejudice against numbers in qualitative analysis is deep rooted, but also misplaced. There may be much to be gained from acknowledging a numerical dimension to analysis, not least in the identification of patterns. Pursuing this line of thought suggests that fuller forms of coding may after all have a heuristic value in generating theory, whether or not they are also required for its validation.

THEORETICAL SATURATION

"Enough is as good as a feast." "Less is more." These aphorisms could have been written just to advise when to stop coding in grounded theory. When a category has become theoretically saturated in grounded theory, then the coding for that category can be brought to a conclusion. Glaser and Strauss suggest that categories are saturated when no further properties or relationships of note are generated by the data: "saturation means that no additional data are being found whereby the sociologist can develop the properties of the category" (Glaser & Strauss, 1967, p.61). It is the capacity of the data to generate new ideas that is exhausted here, and not the accumulation of evidence to support those ideas.

The term "saturation" in this context may seem a rather unfortunate metaphorical extension in which (perhaps more with an eye on academic legitimation than linguistic consistency) the original meaning of saturation has been extended somewhat inappropriately. Saturation has connotations of completion, as when clothes are completely soaked; or even of excess, as when an area is bombed to oblivion. A substance is saturated when it ca hold no more of the material being absorbed. But this claim may seem paradoxical if we extend it from the generation of ideas to the accumulation of evidence. It implies only that the process of theoretical generation is complete—but the confirmation of the ideas so generated is incomplete, since it leaves their verification to further research.

Rather than being described as "saturated," categories produced through and supported only by partial coding might be better described as "suggestive"—as indeed is implied in the reference of Glaser and Strauss to "suggesting" theory. This conveys better their tentative status, though it does not carry the overtones of "systematic" comparison that Glaser and Strauss wish to imply. Or perhaps this middle ground of partial coding might be best represented if the resulting categories (or theories) were described as "indicative"—since this is stronger than "suggestive" (just as the indicative mood is stronger than the subjective) in claiming something to be the case. But even this is perhaps too strong a claim—as Glaser and Strauss seem to mean only that we should be systematic in the procedures we use to generate categories (and their properties), rather than systematic in the accumulation of evidence in their support.

The term "saturation" undoubtedly has a metaphorical resonance that has appealed to many researchers, which may be unfortunate if it is easily misunderstood to imply that data sources have been systematically exhausted. It may be more appropriate to refer to category "sufficiency" rather than "saturation" as the appropriate point at which partial coding can be stopped. Theoretical "sufficiency" would then refer to the stage at which categories seem to cope adequately with new data without requiring continual extensions and modifications. "Saturation," on the other hand seems to imply that the process of generating categories (and their properties and relations) has been exhaustive rather than merely "good enough."

Even the idea of theoretical sufficiency leaves us with the problem of judging when this desirable state has been reached. Perhaps it is no coincidence that Glaser and Strauss in defining theoretical saturation refer to the present tense, claiming only that "no new data are being found" which might serve as a catalyst for further conceptualization. The problem here lies in the unexpected, which may be lying in wait for us just around the next empirical corner. Therefore, a decision not to collect further data can be no more than a guess (albeit more or less well grounded) that such an investment is no longer worth the trouble given the likely (theoretical) reward. We certainly cannot predict accurately whether the very next round of data collection (or even a further trawl through our current dataset) might throw up something that suggests an important modification or even a new perspective.

This continuing unpredictability is recognized by Glaser and Strauss in other contexts:

> When generation of theory is the aim, however, one is constantly alert to emergent perspectives what will change and help develop his theory. These perspectives can easily occur even on the final day of study or when the manuscript is reviewed in page proof: so the published word is not the final one, but only a pause in the never-ending process of generating theory (Glaser & Strauss, 1967, p.40).

Glaser and Strauss are obviously referring here not to new data, but to new interpretations of the available data. However, if interpretation is so volatile, then it also becomes more difficult to assess the point at which categories are developed so completely as to justify a decision to stop collecting data. There must always be the chance that the next batch of data may challenge and perhaps undermine the whole conceptual edifice. Life is full of examples where long-standing assumptions are dissipated in the light of new revelations—one need only think of a wife's discovery of that illicit hotel receipt in her husband's pocket. Indeed, Glaser and Strauss do acknowledge (in a footnote) that the "small chance" that additional data can "explode" an established analytic framework is actually "especially characteristic" of qualitative research (1967, p.73).

On the other hand, in grounded theory this potential for unpredictability may be reduced and made more manageable partly through the procedures that the analyst adopts in collecting data. As analysis develops, it becomes more focused and the procedures for sampling and data collection become more circumscribed. Otherwise, the constant comparison method would entail a sampling and data collection sequence of quite substantial proportions—at least from the point of view of the ethnographer who wants to become fully attuned to the social setting. For example, Glaser and Strauss suggest at one point that it is "not too difficult" to sample 40 groups "on the basis of a defined set of categories and hypotheses" (Glaser & Strauss, 1967, p.70). Theoretical sampling directs the process of collecting data to the selected categories and so reduces the mass of data collected to reasonable proportions.

By contrast, Glaser and Strauss can be quite dismissive of single case studies:

> saturation can never be attained by studying one incident in one group. What is gained by studying one group is at most the discovery of some basic categories and a few of their properties. . . . For example, from studying one incident in one group we might discover that an important property of nursing students' perspectives about course work is their assessment of the differential importance of certain kinds of course work to the faculty; but this discovery tells us almost nothing. To find out such properties as when and how an assessment is made and shared, who is aware of given assessments and with what consequences for the students, the faculty, the school, and the patients who the students nurse, dozens and dozens of situations in many diverse groups must be observed and analyzed comparatively (Glaser & Strauss, 1967, p.62).

An obvious question arises about how "dozens and dozens" of situations can be observed satisfactorily—at least if observation requires the establishment of rapport, a thorough knowledge of the site, sensitivity to context, and time to observe social processes at work. Elsewhere, the authors emphasize the practical value and relevance of grounded theory because of its fit with lay perceptions and experience. But how is this fit to be achieved in so many different settings and groups? Theoretical sampling may give direction to observation, as compared with studies that are unfocused efforts to characterize "everything that's going on." But the real concern here is not with the amount of data being collected, so much as its quality.

Glaser and Strauss acknowledge that establishing rapport is "time-consuming" and that with a succession of sites to sample, time is in short supply. Therefore, they dispense with rapport in favor of what might be called (unkindly, I admit) a "smash and grab" strategy:

> . . . establishing rapport is often not necessary. In later stages of the research, when sampling many comparative groups quickly for data on a few categories, the sociologist may obtain his data in a few minutes or half a day without the people he talks with, overhears or observes recognizing his purpose. He may obtain his data before being shooed off the premises for interfering with current activities; and he may obtain his data clandestinely in order to get it quickly, without explanations, or to be allowed [sic] to get it at all (Glaser & Strauss, 1967, p.75).

Setting aside any ethical issues raised by this passage, we can reasonably contrast this strategy for data collection (justified by the need for comparison) with other strategies that demand weeks, months or even years of slow, careful, and painstaking observation. In so emphasizing breadth of comparison, does grounded theory prejudice claims to the achievement of adequate grounding in terms of quality and depth? More to the point—given the concern in grounded theory with generating theoretical insights rather than the accumulation of evidence—might the predictability of new data and its failure to generate further insights not simply reflect the increasingly focused and limited methods of observation through which it is generated?

This brings us back to the point raised by Merton, noted in the previous chapter. Is the accumulation of evidence not itself a basis on which ideas can be generated? For example, if we want to base our analysis on emerging patterns in the data (of which more below) then might the systematic and full coding of all available data not be preferable to a partial coding which, in effect, stops once we seem to have run out of new ideas?

AMBIGUITIES IN CODING

This issue of when to stop coding reflects a more basic ambiguity in how coding is conceived and presented in grounded theory. On the one hand, coding is conceived as an analytic process, capturing insights and identifying and expressing theoretical themes. This is the coding of the scribbled note in the margin and the amplifying memo: a lightening conductor for generating and relating ideas. It is coding as "an adventure, almost a game" (Richards & Richards, 1991, p.44). On the other hand, we have coding that is systematic and categories that reach saturation through the exhaustion of theoretical possibilities. Thus coding can generate theory that is nevertheless grounded, even though the procedures for the former do not seem entirely inconsistent with the claims of the latter.

This basic ambiguity in grounded theory has been replicated (and if anything, amplified) rather than resolved by the software applications that now support grounded theory. Here, for example, are two (contradictory) claims in a text newly arrived on my desk, and as fair an example as any of the "state of the art" in qualitative analysis:

> In one sense, a code is always a simple label referring to a piece of text, like a Dewy decimal code refers to the location of a book in a library. This sense of a device is captured in the use of the term "indexing" in Nudist, which has been carefully chosen to suggest the filing of a pointer to text, and thus to discourage researchers from overburdening the pointer with theoretical significance (Fisher, 1997, p.69).

> . . . indexing within Nudist is conceived as a "theoretical process," rather than a "code-first, think-next" approach (Fisher, 1997, p.72).

Here we have coding (indexing) presented as "the filing of a pointer to text" divested (or perhaps devoid?) of theoretical significance; but also as a "theoretical process" in which coding does not precede thinking (theorizing) but actually requires it. Mike Fisher has spent a long time using, studying, and writing about software applications for qualitative analysis. If he is confused, then woe betide the rest of us!

In fact, this ambiguity over coding reappears time and again, despite some valiant efforts to exorcise it by distinguishing sharply between the two senses that are confused above. Tom and Lynn Richards (developers of Nud.ist) tried to excise the problem by avoiding altogether "that ambiguous term coding" (1991, p.45). On the one hand, they see coding as a process (as in grounded theory) whereby the researcher "takes off" from the data—that is, "explores, examines, theorizes about emerging ideas." On the other hand, they describe coding as "a process not unlike postcoding for quantitative analysis" whereby the researcher attaches "the same tag to all the stuff on one subject" so they can find all the "stuff" with a given tag (Richards & Richards, 1991, p.44).

These coding procedures are presented as involving the pursuit of different goals—to generate theory, on the one hand, as against "to collect extracts in a manageably few categories for efficient retrieval," on the other (Richards & Richards, 1991, p.44). Theory generation (it is argued) requires a light touch—what I have described as partial coding. Theory testing requires the labor of applying all codes systematically to all the evidence so that they retrieve all the instances that fit the theory—or what I have called full coding.

Having distinguished these two senses of coding, the developers of Nudist eschew the term and opt for another: that of "indexing" data. However, the confusion simply refuses to be banished, surfacing again immediately in their discussion of indexing. Thus indexing is introduced as one of the "clerical" tasks—of recording, storing and indexing data—that can be managed by software such as Nud.ist. Here indexing seems to mean something akin to a book index—merely a means of locating (and retrieving) relevant data. But in the same paragraph, we learn that "exploration and theory development" involves an interplay between the original text and "the conceptual structure used for and created by its exploration"—which refers, presumably, the indexing system. Thus "indexing" can refer either to a simple clerical procedure for tagging parts of the text or to a complex conceptual process for theorizing about the data. This is not so surprising (and indeed can be seen as a strength), since Nud.ist was designed to support either or both processes. Nevertheless it leaves us in a muddle about what coding (or indexing) may involve.

FACTUAL CODING AND REFERENTIAL CODING

In a later text, Richards and Richards (1995, p.84) distinguish between the coding of "factual" and "referential" categories. Factual categories refer to attributes of the data, such as the background characteristics (such as age, gender, etc.) of the respondent. Referential categories refer to the location of parts of the data; that is, they contain references to where a particular topic (represented by the category) appears in the data (Fig. 6.1).

Inspired by the above distinction, but also reinterpreting it, Seidel and Kelle (1995) also identify two modes of coding textual data that they also describe as "denoting a text passage" and "denoting a fact" (Fig. 6.2). However, they go on to note that "codes themselves are neither factual nor referential," since it is really a question not of what codes are but of how they are used. They also suggest that "coding always has both functions" as there are no factual codes that make no reference at all to data. However, one function may be accorded priority over the other depending on the particular requirements of the research (Seidel & Kelle, 1995, p.53).

Thus Seidel and Kelle retain the idea of "referential" codes as referring to

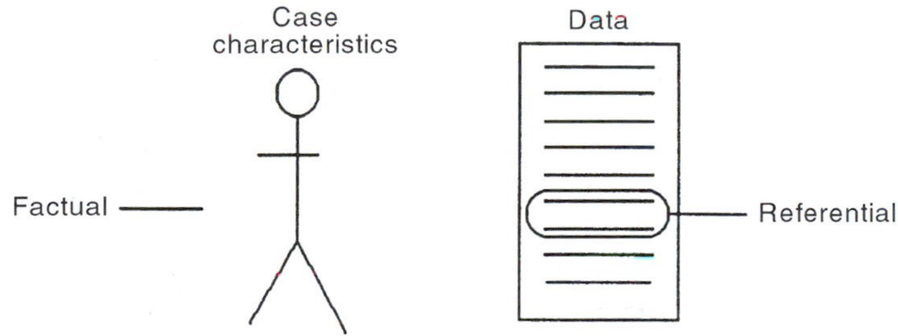

FIGURE 6.1 Factual and referential categories (derived from Richards & Richards).

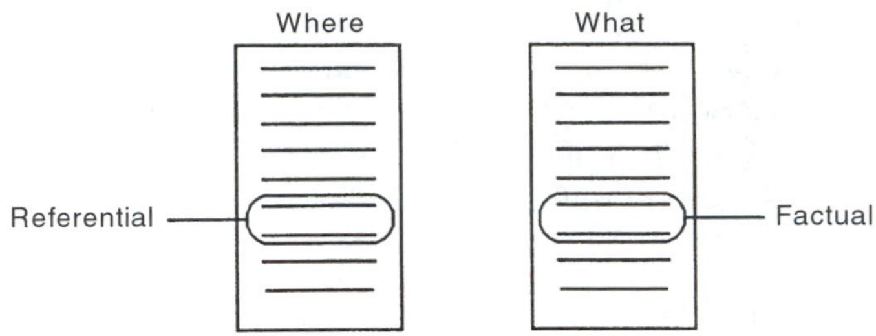

FIGURE 6.2 Factual and referential categories (derived from Seidel & Kelle).

locations in the text, but apply the term "factual" coding to the content of data as well as background case characteristics.

Having made this distinction, Seidel and Kelle express concern about what happens when these two modes (or functions) of coding are confused. Noting the "fascinating new possibilities" opened up by computer software for qualitative analysis, they also note "specific dangers":

> . . . the same technical tool can be used for two totally different research strategies which employ codes differently: as indexes for text segments that are coded and retrieved by an electronic "cut-and-paste" device on the one hand, and as representations of facts contained in the raw data on the other. Therefore analysts can—without realizing—confuse the two modes of coding: they can involuntarily switch from using the referential function of codes (that means from collecting text segments that refer in a broad and general way to a number of somewhat vaguely defined concepts) to treating codes as if they were representations of factual information (Seidel & Kelle, 1995, p.59).

They see two dangers in this confusion. One danger is of "losing the phenomenon by reifying the codes," whereby the researcher treats referential codes as though they were factual, and forgets the data to which they refer. This is illegitimate since:

> . . . there is only a loose coupling between a code and a piece of data instead of a well-defined relation between a code and a phenomenon, since the code was not attached to denote a certain discrete event, incident or fact, but only to inform the analyst that there is interesting information contained in a certain text segment, related to a topic represented by a code (Seidel & Kelle, 1995, p.59).

The second danger is of "losing the context"—of "counting the grass" and not "seeing the grazing"—that is, of accumulating and comparing codes, but not recognizing the meanings of the text segments referred to, which may depend on the contexts in which they appear.

The danger, in short, is that codes once formulated will establish a life of their own, independent of the data to which they refer. The analyst starts to trade in codes, without sufficient regard to the evidence that can be accessed through these codes and invests them with meaning. This raises again the more general issue of the role that coding should play in the production of evidence to support ideas.

HEURISTIC AND REPRESENTATIONAL FUNCTIONS

Put another way, we can ask whether coding serves a "representational" or "heuristic" function? In "representational" coding, the "codes serve as representations of the investigated phenomenon" (Seidel & Kelle 1995, p.53). In "heuristic" coding, codes are "imprecise and vague" (Seidel & Kelle 1995, p.58), but they need to be so, it is suggested, in service of "a methodology of discovery":

> Here, in contrast to a methodology of hypothesis testing, the researcher must not restrict the scope of the investigation in advance by determining precise categories, since the goal is not to recover certain already known phenomena in the empirical field but to discover new ones (Seidel & Kelle, 1995, p.58).

Thus in "representational" coding, the code stands for (or "denominates") the content of the data, while in "heuristic" coding, the code serves as a "device for discovery" (cf. Fig. 6.3).

Seidel and Kelle suggest that representational coding characterizes hypothetico-deductive research strategies that impose quite stringent analytic requirements. Any coding in hypothetico-deductive research should be systematic, consistent, inclusive and exhaustive—with coding of every instance of a category (or, in other words, what I called full coding). For this to be possible, the relevant variables and values must be determined before the final coding begins. By contrast, heuristic coding is regarded as a characteristic of interpretive inquiry where the meanings of actors (and action) cannot be established

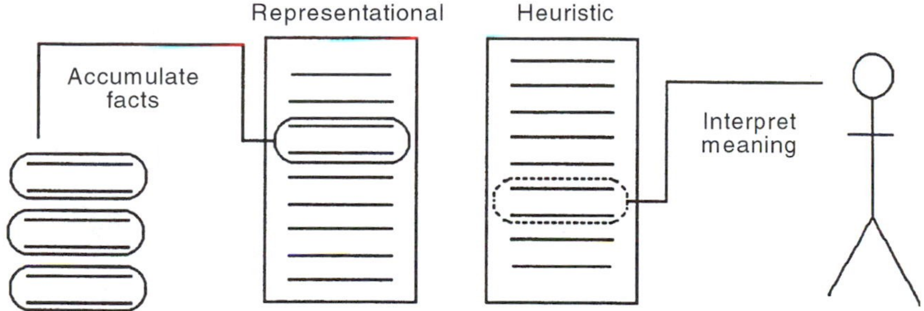

FIGURE 6.3 Representational and heuristic functions of coding.

in advance of the research. This requires the researcher "to go out and look at how people perceive and interpret the world themselves" (Seidel & Kelle, 1995, p.55). Here coding is used to identify all the relevant data germane to a particular theme so that the data can be reordered to allow comparisons to be made and meaningful patterns to emerge. The aim is to generate ideas rather than test them.

It is not clear quite how the factual/referential distinction figures in this contrast between the representational and heuristic functions of coding. It may seem plausible to argue that representational aim requires factual coding while heuristic ambitions can be satisfied by referential coding. If so, we seem to slip effortlessly from a distinction which began as a subtle and pragmatic matter of emphasis (on factual or referential coding) to one which becomes a full clash of methodologies (hypothetico-deductive testing vs interpretive discovery).

This analytic slippage may occur because the factual/referential distinction was never very clear in the first place. Suppose we return to the point made by Seidel and Kelle that coding is always both factual and referential. If so, then this implies that coding in interpretive research is not just referential. Codes never "just" indicate (or signpost) the relevance of a text segment to a particular theme. By coding that segment of text as relevant to that theme, we already conceptualize the data in a particular way—that is, we ascribe particular meanings to the text. The process of discovering meaning does not follow after coding but is initiated through it. This is true even where codes are simply numbers or marks assigned in the margin to passages of text. These numbers or marks are not entirely innocent of meaning—they are codes that already signify implicit or anticipated categories that are thereby assigned to the text. When we assign a category, we are always saying something (not matter how obscure) about the text that distinguishes it from other parts of the data. Of course, this conceptualization may be very vague and imprecise—but that does not reduce codes to a referential function.

The factual side of the factual/referential distinction is also suspect. It hardly matters in this context whether a code refers to a particular part of the data or the code refers to a document (or a case) as a whole. The important point concerns the claim that codes can denominate phenomena in a way that is "precise, systematic, consistent, inclusive, and exhaustive." It may be natural to think of codes in this way—but not the categories they stand for. We need only recall here the problems of the classic model of categorization discussed in the previous chapter.

Let us imagine a very simple example. Suppose we code the following passage for the word "duck."

> I took the children to the park to feed the ducks. I wanted to duck out of it but my wife wouldn't let me. I told her I had a book to write but it was water of a duck's back. Don't get me wrong—she's very ducky. As we passed the cricket ground we

> stopped to watch for a while. We saw my cousin just going into bat—he was out for a duck. His successor hit the ball over our way and I told the children to duck. When we got to the park I ducked the children in the pond.

We could attach a code "1" for each sentence containing the text "duck." That would give us the following:

1 I took the children to the park to feed the ducks.
1 I wanted to duck out of it but my wife wouldn't let me.
1 I told her I had a book to write but it was water of a duck's back.
1 Don't get me wrong—she's very ducky.
0 As we passed the cricket ground we stopped to watch for a while.
1 We saw my cousin just going into bat—he was out for a duck.
1 His successor hit the ball over our way and I told the children to duck.
1 When we got to the park I ducked the children in the pond.

What is the factual content of this coding? The code of "1" tells us that the target text appears in each of these sentences. We could conclude that the text "duck" appears in 7 of the 8 sentences. This coding is "precise, systematic, consistent, inclusive and exhaustive"—and meaningless. Although the target text "duck" appears in each of these sentences, it has a different sense in each of them. Each fact has to be interpreted—and following Lakoff, we could suggest that its interpretation depends on the relevant idealized cognitive models. When we consider what they mean, the facts about these ducks prove fuzzy.

Even if we ignore meanings and stick with the supposedly "hard" evidence of textual content, we have to recognize that we have produced (and not just collected) our facts. Consider the examples where the text "duck" appears as part of a word: "ducks," "duck's," "ducky," and "ducked." In effect, we have produced seven sentences including the text "duck" only because we have followed an implicit rule that allows parts of words to be included in our coding. If we had followed a different rule—say confining our coding only to the appearance of "duck" as a whole word—we would have produced a different result. Just to underline the point, suppose we have to code the following continuation:

> The pond was covered with weed—"uck" said the children. But on the whole I think they conduckted themselves well.

With the first sentence, we have to decide whether to include the "duck," which is formed from "wee<u>d uck</u> said" if we ignore punctuation. With the second sentence, we have to decide whether to include the "duck" that appears in the misspelling of "con<u>duck</u>ted"—though certainly no one intended a "duck" to be there! In other words, we have to decide what constitutes the facts, even with respect to the appearance of a term in the text. We have to consider where the boundaries of a category lie, what underlying conceptualizations it expresses, or what rules and procedures (if any) we can devise to govern membership.

This seems simple and precise, but only so long as we confine ourselves to thinking about "good exemplars"—such as a "prototype" instance of a textual occurrence of a "duck"—in mind. As we have seen, however, Lakoff would argue that these exemplars are invested with significance only through the implicit reference to idealized cognitive models. If, on the other hand, we claim that our perception (and production) of facts is governed by rules, it is also difficult to claim that they are sufficiently "precise, systematic, consistent, inclusive and exhaustive" as there is usually (always?) room for some ambiguity in the application of such rules to a particular case. As Harnad argues, even with a rule-based approach we can only converge on an approximation that discriminates among confusable alternatives.

These ducks will have served their purpose if they have illustrated the potential fuzziness of factual coding, implicit in either the rules governing the production of facts or in the categories through which they are interpreted. The basic point is that factual coding involves conceptualization, just as I suggested was the case with referential coding. The distinction between them is therefore only a matter of degree—as Seidel and Ulle argued in the first place! Thus referential codes must be factual to some degree (in the sense of making some claims about the data) and factual codes must be referential to some degree (in the sense of implying rules which refer to the data). Unfortunately this leaves the ambiguity over the factual or heuristic functions of coding to test or discover theory unresolved.

REDUCTION AND COMPLICATION

Another effort to resolve this ambiguity over coding can be found in the distinction made by Coffey and Atkinson (1996, pp.28–9) between coding for "data simplification or reduction" and coding for "data complication" (cf. Fig. 6.4). This again contrasts coding as conceived in "code-and-retrieve" routines with coding as conceived in grounded theory. The "code-and-retrieve" approach allows data to be "reduced" to "equivalence classes and categories" used to retrieve all the data sharing a common code:

> Such code-and-retrieve procedures can be used to treat the data in quasi-quantitative ways by, for example, aggregating instances, mapping their incidence, and measuring the relative incidence of different codes (Coffey & Atkinson, 1996, p.28).

This form of coding is only "quasi-quantitative" because "we are not merely counting" but also "attaching codes as a way of identifying and reordering data, allowing the data to be thought about in new and different ways" (Cof-

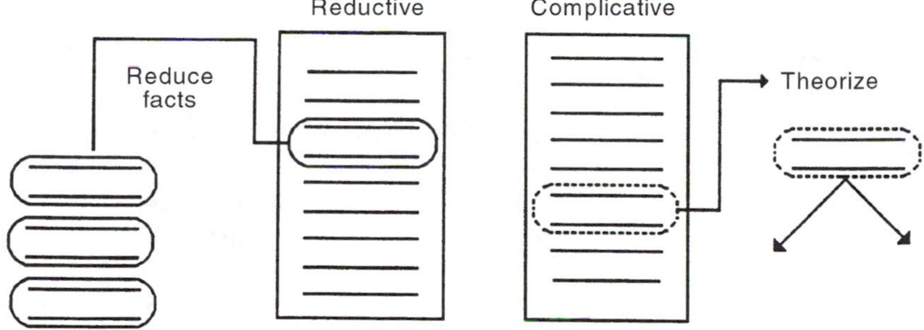

FIGURE 6.4 Reductive and complicative forms of coding.

fey & Atkinson, 1996, p.29). On the other hand, we have coding as "data complication," as in grounded theory:

> Coding need not be viewed simply as reducing data to some general, common denominators. Rather, it can be used to expand, transform and reconceptualize data, opening up diverse analytic possibilities (Coffey & Atkinson, 1996, p.29).

The orientation is not toward simplification and reduction, but toward interrogation and speculation:

> Such data complication is not used to retrieve and to aggregate instances to a restricted number of categories; rather, it is intended to expand the conceptual frameworks and dimensions for analysis. Coding here is actually about going beyond the data, thinking creatively with the data, asking the data questions, and generating theories and frameworks (Coffey & Atkinson, 1996, p.29)

Having made the distinction, Coffey and Atkinson also suggest that coding in practice is usually a mixture of both. Coding is generally used both to organize data into simpler general categories and to formulate new questions and interpretations. So there is a balance to be struck between reduction and complication, and the researcher "should try to ensure that coding does not lose more than is gained" (Coffey & Atkinson, 1996, p.30). In principle, though, this balance is weighted toward data complication:

> It is especially important to avoid the use of coding merely to apply simple and deterministic labels to the data. Data reduction or simplification of that sort is not the main analytic purpose of qualitative coding. Coding should be thought of as essentially heuristic, providing ways of interacting with and thinking about the data (Coffey & Atkinson, 1996, p.30).

In this view, then, coding is "essentially heuristic." Coffey and Atkinson cite, as an example, Tesch's view of coding as "a means of providing new contexts for viewing and analyzing data." With even more approval, they cite grounded theory as an example, notably the argument that:

> . . . coding is much more than simply giving categories to data; it is also about conceptualizing the data, raising questions, providing provisional answers about the relationships among and within the data, and discovering the data (Coffey & Atkinson, 1996, p.31).

In other words, coding is aimed at thinking about the data and opening up inquiry.

While these admonitions against a simple and reductive approach to coding are hopefully well taken, they may also seem puzzling. After all, even a code-and-retrieve approach can be described as essentially heuristic, allowing data "to be thought about in new and different ways." If code-and-retrieve is the dragon to be slain, it seems an elusive one. If even a process of simple, reductive coding requires reordering and reconceptualizing data, then the contrast between data reduction and data complication becomes much less clear-cut. All the more so, if we recognize that the complications of grounded theory are supposed to be achieved at least in part through data reduction. Data reduction and complication then become, not alternatives, but rather two sides of the same coin.

CONFLICTING CLAIMS

The reader may have noted a common style to these discussions of coding, which involves setting up a distinction between two methods, and then advancing arguments which tend to favor one method over the other. Thus Seidel and Kelle plainly favor the heuristic coding of interpretive inquiry, and indeed they even question whether representational coding is feasible in qualitative analysis. Similarly, Coffey and Atkinson clearly favor complicative rather than reductive coding. In each case we end up with one way of doing analysis (and one way of not doing it). But the options presented depend on distinctions that the authors struggle to sustain throughout the details of their argument.

Nevertheless, the general thrust of these arguments is in line with the "logic of discovery" of grounded theory. Questions concerning the accumulation and use of evidence in coding are conveniently sidelined by the dialectic of opposing strategies, which allows the factual and reductive aspects of coding to play very much second fiddle to its heuristic value. This repeats the claims of grounded theory, but without really making any clearer how coding can reconcile the potentially conflicting claims of theorizing and grounding.

As Fisher notes, an interesting side issue in these discussions concerns the question of "whether coding is seen as an integral part of the analytical process or as a precursor of analysis" (1997, p.71). Fisher suggests that the answer is clear in principle but less so in practice. In principle, it is recognized that "unless coding is regarded as part of analysis, researchers are reduced to spotting predicted patterns in the data." But in practice, much of the software "implies a separate process of coding which must precede analysis using codes" (Fisher, 1997, p.71). While the principled answer stresses the interplay of coding and analysis, the practical answer implies their separation into successive stages—the "code-first think-later" approach. This is reinforced by the presentation of coding as a rather dreary clerical task that is counterposed (as preparatory) to the exciting tasks of analytic exploration and discovery. The potential user is given conflicting messages about their epistemological responsibility as researchers and their practical aims as users of software. The former requires them to think first and code later; the latter invites them to code first and think later.

In this respect, the problem for software developers and users espousing grounded theory is that partial coding may tend to discount the very powerful tools that computer applications can offer. Computers are excellent machines for searching databases and counting examples. This technology encourages both developers and users to think in terms of coding data fully so that all relevant instances can be retrieved, compared, and perhaps even counted. None of this seems required by partial coding—and indeed it is explicitly rejected time and again by Glaser and Strauss as exemplifying the "verificational" approach. Thus we have a potential mismatch between the methods of coding supported by software and the canons of grounded theory—which seems somewhat paradoxical given the popularity of the latter as a rationale for the development and use of software applications.

CODING OR CATEGORIZING

How can we resolve the persistent muddle over coding? A first step, in my view, would be to ditch the term "coding" altogether (and the underlying cognitive model that invests it with meaning). I suggested earlier that it was used in grounded theory to legitimate qualitative research. However, the metaphorical and metonymic extensions involved in applying the term to qualitative analysis are quite misleading.

There can surely be no term more inappropriate than "coding" to describe a procedure that is purportedly central to qualitative analysis. The term is an example of a metaphorical extension, in which an original meaning has been extended to a new—but in this case, entirely inappropriate—context. We refer to legal statutes that are organised to eliminate inconsistencies and overlap as "codes." More commonly, we think of "encoding" as method of translating language into a secret set of symbols, whose meaning can only be divulged by "cracking the code." Nowadays we also have computer languages translated into "machine code," a set of arcane symbols instructing our computers how to process information. These meanings have connotations of consistency, logic, and mechanical translation that seem quite at odds with the conceptual and iterative processes of qualitative analysis.

Coding also exemplifies a metonymic extension, as it substitutes part of a process for the whole. The part concerned is the use of codes as brief symbols (names or labels) in the place of categories, to facilitate identification and retrieval—as, for example, if you were to write "Satpm" to refer to an event that you have assigned to the category "Saturday afternoon." This substitution of a category by an identifying code name may be useful in simplifying the mechanical processes of handling or presenting data. However, the extension of the term for what is an essentially mechanical and arbitrary process to the diffuse tasks of organizing and analyzing qualitative data seems singularly inappropriate. The codes themselves have no meaning aside from the codebook which specifies the symbols to be used in translating the source into code. In the pre-coded questions used in survey research, the numerical codes add nothing to the meaningful interpretation of data—they merely identify responses efficiently and economically for subsequent processing.

Much of the discussion noted above has been directed at distancing qualitative analysis from a mechanical, meaningless and mindless process of attaching labels to text segments. But that is precisely what coding connotes. So long as qualitative researchers think in terms of coding data, the spectre of reductive and representational approaches will probably continue to haunt their efforts to produce meaningful interpretations of their data. These fears may be unfounded, I have argued, insofar as qualitative analysis simply cannot proceed even in the crudest form without involving (though perhaps without confronting) the problems of conceptualizing data. But so long as the language (reinforced by the software) implies otherwise, this is a battle that researchers will always have to fight.

In place of coding, my first inclination (though with some reservations) is to suggest the term "categorizing" to describe a process which, after all, is usually taken to involve the assignation of categories to parts of the data. Indeed, it is precisely because it involves categorization that we cannot reduce this process to the mechanical procedures implied by the term "coding." One virtue of describing the process as "categorizing" is that it opens the door to a much wider interpretation of what this involves. In a previous chapter, I noted some of the separate problems—such as identification and discrimination, or category and cue validity—that have been explored in linguistic and psychological research. If interpretive inquiry is to be orientated to and perhaps modeled on common sense, as some advocate, then it should surely become more reflective, more critical, and more informed about the character (and limitations) of common sense thinking. In light of cognitive research, a common sense approach implies some sensitivity to the problems of fuzzy boundaries and graded membership, the role of prototypes, and the import of idealized cognitive models. Categorization raises all of these issues—while coding raises none of them. To my mind, categorization poses new challenges and prospects, while use of the term "coding" requires energy and effort to be squandered on fighting old battles.

My reservations are concerned with the limitations of categorizing, which should not be taken to constitute the whole analytic process. The "container" metaphor which underpins categorization does not exhaust the possibilities of human inquiry. As we see, there are other, equally productive, modes of thought—such as the "source-path-goal" schema (Lakoff, 1987, p.278) of narrative analysis—which can open up different lines of inquiry (cf. Coffey & Atkinson, 1996).

INDEXING

Earlier I expressed some scepticism over name substitution as a means of resolving ambiguity, and suggested that substituting the term "indexing" for that of "coding" did not clarify the underlying issue of what coding requires. So why should substituting the term "categorizing" for "coding" make any difference? I do think names matter, if only because of the underlying cognitive models that motivate their use. The problem with indexing is that it extends the everyday cognitive models of book or library indexing to the analytic process. These models of indexing suggest a process which is more or less mechanical. The book index is formed by the frequency which terms occur in the text. The library index is shaped by the classification scheme which determines where items are stacked. For different reasons, neither are subject to easy revision. Once it is in print, you cannot tear out the index of a book and start again—reindexing must await a new edition. The library classification scheme also endures, entrenched in the computer system, the professional training of librarians, and the need for consistency across libraries. This durability lends to the process of indexing a kind of semi-permanent gloss, sheened perhaps by the apparent authority of expertise and tradition. Applied to qualitative research, this implies that indexing can provide a semi-permanent platform on which to explore the commonalities, links and patterns in the data.

Lakoff of course stresses that such cognitive models are idealized—they bear only an approximate resemblance to the reality they purport to capture. Indexing a book in practice requires conceptual decisions that are far from routine. Is a term (concept) important enough to merit a place in the index? Should the index be "flat" or should some terms appear as main entries and others as subdivisions?

The same point applies to library classification. Even with an established classification scheme, there may be important and difficult judgments to be made about the classification a text should be assigned. I know this as a bitter truth, for librarians barely recognize my own subject (social policy) in their classification schemes, leaving students of the subject to hunt down relevant texts in many disparate locations. This exemplifies another important point—that there is no single satisfactory method of indexing the material. Any classification scheme, while not arbitrary, is not objective in the sense of reflecting a uniformly agreed and universal schema. It is a product of our understanding about how books can be best organized—for some practical purpose, such as locating them easily—and such understandings do differ according to the interests of the user. If this were not so, I would not waste a few hours every year or so reorganizing the books on my own shelves under some new schema (alphabetical order, subject-matter, current relevance) which I foolishly imagine will somehow solve the perennial problem of readily locating a needed text.

Those adopting the term "indexing" themselves point out significant differences between this term and the process of categorizing as it is typically understood in qualitative research (Richards & Richards, 1995, pp.80–1). In contrast to library indexes, which provide a "firm framework for storing homogeneous contents" a category system offers a "flexible container for complex contents" (Richards & Richards, 1995, p.81). One may doubt one aspect of this contrast between library indexes and category systems, for though books may be homogeneous in appearance, they are heterogeneous in content. The other aspect contrasts the firm structure of library indexes with the flexible structure of category systems, which "must have not only a structure but also ways of adapting and reviewing that structure" (Richards & Richards 1995, p.81). Compared with a library index, a category system is more tentative and provisional. Moreover, the categories constructed and employed in the research have to be examined, documented, and authenticated—they cannot simply be taken for granted.

Of course, we may want to organize and manage our data in much the same way as we index a book or classify the contents of a library. My point is not to dispute the practical requirements so much as the underlying idealization that informs them. One of the first things we have to decide is how we are going to store data—and that decision affects the way we retrieve it. Suppose we simply store data under case heads in a particular order, say alphabetically. Then that decision already reflects two important judgments—about how the data can be divided into cases, and how those cases should be ordered. Such judgments may be vital in determining the shape of the research. The process of storing and managing the data already involves, in other words, significant decisions about categorizing and ordering (no matter how obvious or self-evident they may seem to the researcher) that the mechanical connotations of the term "indexing" (whatever the explicit intention of its authors) tend to discount.

To claim that categorical judgments are involved in even simple procedures for filing and storing data does not mean that these must be subject to perpetual revision. It does mean, though, that their provisional character should be recognized and made explicit. However, we can still invest more confidence in some judgments than in others. We can envisage a confidence continuum from those judgments in which we have a great deal of confidence (for example, that the sun will rise tomorrow) to those where we may be less certain (for example, that we will be around to see it). Our confidence in each case reflects our background understandings. With regard to the sun, we know that in the normal course of events stars die, but ours is not expected to expire in the near future (at least in human terms), while the chances of a cosmic accident are slight. With regard to our own fate, our (lack of) confidence reflects our limited knowledge of the age-related vagaries of sudden illness and disease and the ever-present chances of accidental death. Similarly, there may be some judgments that we make in the course of research in which we have a great deal of confidence, while other judgments may be less certain. Much may depend on the context in which judgments are made and our purposes in making them. The process of categorizing may not be "mechanical" but that does not mean that it is entirely arbitrary.

CONFIDENCE IN CATEGORIES

Coding is used to "ground" theory in data. But how can we be confident that our categories are sufficiently grounded—that they are indeed "indicated" by the data?

One answer may lie partly in the distinction we noted in the previous chapter between basic-level and superordinate or subordinate categories. Basic-level categories, you may recall, are those we learn most quickly, identify most readily, interact with most frequently, and communicate most efficiently. Take a simple, directly comprehended statement such as "the cat sat on the mat." The "cat" and the "mat" are basic-level categories that we understand directly. We know what a "cat" is, for example—at any rate in this context, which the reference to "mat" defines as domestic. The same is true of the "mat." These are examples of the basic-level categories which we acquire through our early (perceptual) experience of and (physical) interaction with the world. Though the term "on" is more complex—Lakoff (1987, p.293) suggests it is composed of three "kinesthic" image schemas "above," "contact" and "support"—we also have a direct understanding of what it is to be "sat on" (or to "sit on" something) through our own bodily experience.

Because of this direct understanding, we may have more confidence in categorization involving such basic-level categories than in more superordinate or subordinate levels of categorization. We may be sure that the "cat" is a cat, but less certain whether it is a Persian, Siamese, or some other variant of the species. Recalling that elephants and lungfish can occupy the same niche in a cladist evolutionary classification, we may be even less sure how our cat fits into a wider biological taxonomy. Much the same point applies to mats. We can be sure our mat is a mat—if only because a cat is sitting on it—otherwise we might call it a carpet! But suppose we have to categorize the contents of a furniture store, and place our mat in some appropriate category to facilitate stock maintenance or review. In this less familiar and more abstract schema, we might experience more difficulty in categorizing and have less confidence in the result. "Furniture" is not a basic but a superordinate concept—we usually think in images of tables and chairs, and not of in terms of taxonomic categories such as furniture.

To reiterate an earlier point, none of this implies that our categories are symbols that correspond directly to the world. We need not claim that the world is actually full of cats and mats independent of the human beings who experience it (and categorize it) that way. As Lakoff points out, the processes that distinguish basic level concepts—identification, interaction, communication, and organization—are all human activities conditioned by our own capacities for vision, speech, movement, and so on. We see the world "as we see it." But while this precludes a uniform and universal (indeed omnipotent) understanding of the world, it still leaves room for confidence in basic level categorization sufficient for practical purposes.

If we can have greater confidence in basic-level categorization, this may have implications both for where we start to categorize and how we go on to develop our categories. If basic-level categories are those typically located in the middle of a taxonomic hierarchy, then it may make sense to begin analysis by assigning categories at that level. These can then be refined through subordinate categories or integrated through superordinate categories. I do not mean that one begins the analysis only with basic categories. The reason for using a basic-level category in the first place may be because it has a complex internal structure. We may also identify categories at a basic-level as members of superordinate categories at the very outset of the analysis. As I suggested earlier, we can think of our categories as strings that have conceptual connections with other points (categories) rather than as isolated constructs that can be assigned without regard to the overall analysis.

Thus our confidence in coding may reflect the extent to which our categories are embedded in a given conceptual framework. But we also have to consider how far they are also embedded in the data we have produced in the course of the research. If our conceptualizations are to be grounded, do we need to go beyond the light touch of partial coding, and make a more systematic appraisal of the available evidence?

CODING AND COUNTING

As we saw earlier, Glaser and Strauss launched grounded theory with a firm rejection of full categorization. Given the aim of "discovering" ideas, to categorize incidents that prompt no further theoretical elaboration of categories would be a waste of time and effort. The only rationale that requires full categorization (of every incident) is that of verification—that is, to test hypotheses. As we saw earlier, Glaser and Strauss eventually parted ways over this issue, with Glaser insisting still on the merits of partial categorization, while Strauss became more inclined to talk in terms of verifying categories through further confrontation with evidence. Thus with Corbin he suggests that "we return to our data and look for evidence, incidents, and events that support or refute our questions," looking "for evidence in the data to verify our statements of relationships"—or to refute them (Strauss & Corbin, 1990, p.108). Such verifications and refutations are presented in terms of what they add to "variation and depth of understanding" (Strauss & Corbin, 1990, p.109). Nevertheless this not only involves a different language (including that of "verification"—though Strauss and Corbin seem more comfortable with "validation") but it also implies a different approach. Thus Strauss and Corbin argue that

> testing is a crucially important and integral part of grounded theory . . . only that which has repeatedly found to stand up against reality will be built into the theory (Strauss & Corbin, 1990, p.187).

Obviously, it is not possible to look for such repetitions if the relevant evidence has not been categorized in the first place. And where we look for repetitions in evidence, it is but a short step to using numbers in order to identify them.

Despite this hint of revisionism, grounded theory is generally taken to be concerned with conceptualization and so to discount numbers. Indeed, the two processes—of conceptualizing and counting—are often held (or implied) to be mutually exclusive:

> There is little point in counting the pieces of text to which a heuristic code has been applied, since its importance does not lie primarily in its numerical incidence in the data, but rather in the way it illuminates one aspect of interaction (Fisher, 1997, p.70).

This argument is typical of the tendency to dichotomize choices, even where the basic argument is hedged around with qualifiers—such as "little" and "primarily." These hedges hint at a logical flaw in the argument, that tries to draw a firm conclusion (that there is "little point in counting") from a rather tentative premise (that counting is not of "primary" importance). Here the "little point" is perhaps an idiomatic way of suggesting that there is no point in counting. The implicit conclusion that counting is irrelevant is certainly a comforting one, at least to those of us (myself included) who are not comfortable with all those difficult mathematical manipulations that go with it. But is counting so irrelevant to the generation of theory, particularly theory that claims to be grounded? As Merton suggested, might numbers too not serve a heuristic function?

A dichotomy between meaning and numbers runs deep in popular culture. Even though we talk of "figuring out" what things mean, for many that familiar alliance of reasoning and reckoning apparently "counts" for nothing. Numbers and meanings tend to be seen as diametrically opposed. Indeed, we commonly perceive those fixated by numbers as somehow "missing the point." This neatly expressed in the story of the new prisoner who attends an after-dinner theatrical performance, in which one member of the audience after another jumps up to the platform, shouts out a number and the audience collapses in laughter. The secret of their performance lies in a volume in the prison library, entitled *The Best Ten Thousand Jokes,* which of course all the long-term prisoners know by heart. To tell a joke, they just shout out its number. Thus enlightened, the new prisoner leaps to the stage next night and shouts out "5955." Unfortunately, nobody laughs. And why not? Because he didn't tell it right! (Rucker, 1987, p.85).

This story is not just an apocryphal illustration of the cliché: "it's not what you say but the way that you say." It also evokes our suspicion that when numbers are substituted for words, something important may get left out in the process. In a more serious vein, Einstein expressed such doubts with his customary eloquence: "so far as the laws of mathematics refer to reality, they are not certain, and so far as they are certain, they do not refer to reality" (cited in Kosko, 1994, p.7). Ironically, the above story also reflects our love of numbers—so that even jokes can be enumerated (if you do it the right way). Indeed, despite the popular prejudice against numbers—or perhaps mathematicians, described by Barrow as "the people you don't want to meet at cocktail parties" (1993, p.5)—the world of common sense and everyday experience is not an innumerate one. As Barrow observes, "numbers dominate much of our lives in diverse and barely perceptible ways (1993, p.5). We use numbers in all sorts of contexts, from choosing channels and buses to measuring temperature in the street or the oven. Sometimes numbers act as a kind of economical form of coding (properly used in this context): we ask for a ticket for cinema 1, not for the big cinema up the stairs with the more comfortable seats; and we get there by number 11 bus, not the one that goes from one place to another by a particular route. Often we use numbers as a form of measurement so that, for example, we can gauge how far it is to town, how much the fare will cost, and how long it will take us to get there. Numbers are a pervasive part of our everyday life (at least in our "information" societies) no matter how much we like to overlook or deny it. If grounded theory emulates common sense understanding, then using numbers to code or count would be part and parcel of the analytic process. In this respect, it may not depart as much as some imagine from the traditions of qualitative research, since "qualitative researchers are often quite able and proficient counters" (Van Maanen, 1982, p.15)

Just what these numbers tell us is, of course, problematic—"it is a well kept secret of mathematicians that even they don't know what mathematics is" (Barrow, 1993, p.209). I do not propose to try to resolve these problems here (or anywhere else, for that matter)! However, it seems clear that the processes that we associate with categorization, such as identification or discrimination, may often have a numerical aspect. A category does not refer to a single item (which we might give a unique name) but to items in the plural. It refers to a class (or set) of items that we define (somehow) as belonging together. In the classic model, as we have seen, this classification is based on sharing essential features, that serve to demarcate one entity from another. Membership is clear-cut—either an entity has the requisite characteristics or it does not. This suggests that we can categorize on the basis of reasoning alone, whatever we may do with the results. First determine membership, then add up the numbers. In other words, once we have categorized, then we can count. This parallels the logic of grounded theory with its injunction to generate categories first and to verify—and count—evidence later.

However, we have seen that this model may have some limitations, given the role of gradation, of prototypical examples, and of idealized cognitive models. If categories are graded, then degrees of membership can be rated in a process that can be modeled by mathematics. If category members share only a family resemblance, then membership becomes indeterminate but may be modeled through probability and statistics. Even if categories are motivated in terms of idealized cognitive models, this does not preclude the possibility (which unfortunately Lakoff does not consider) that they may have a mathematical dimension. For example, the container metaphor itself suggests the notion of contents that can be enumerated.

There is no need to claim that "number is an objectively existing feature of the world" or that "if any intelligent race does form the idea of discrete objects, we may be certain that they will arrive at just the same numbers that we use" (Rucker, 1987, p.50). On the contrary, numbers themselves can be seen—and appreciated—as a cultural product. For example, sophisticated notational concepts involving place-values (such that the "3" in "341" means "300") developed only in three advanced cultures; while the concept of "zero" was invented in various cultures, with different connotations in each (Barrow, 1993, pp.81–99). While modern mathematics is plainly a sophisticated product of recent cultural development (at least from an evolutionary perspective), some cultures have lacked any systematic notion of number at all and many cultures (perhaps serving as an inspiration for our modern antinumerical qualitative researchers) could only distinguish one from two or many (Barrow, 1993, p.102).

Language (itself a cultural product) predates numeracy.

> Counting appears to have evolved out of the general desire for symbolic representation which language first meets. It is a specific from of this mental symbolism and there is much evidence to suggest that it may have developed from entirely symbolic ritual practices in which notions of pairing, patterns, succession, and harmony were paramount (Barrow, 1993, p.102).

Barrow suggests that counting (as an elementary intuition) must have had its roots "in something more basic than number," perhaps in the shape of "conscious notions of distinction" such as male and female, high and low, and the like. However, Barrow adds that in terms of selective advantage, we must have been able "to distinguish different states of the world at some level of accuracy and to have a means of noticing if quantities altered" (Barrow 1993, p.103).

The recent evidence of teaching language to chimps suggests that some sense of number is deeply rooted:

> . . . this research has convincingly demonstrated that chimps understand several important concepts, including numbers, how to add and subtract, the nature of basic relations (such as "is bigger than," "is the same as," and "is on top of"), how to ask for specific objects (mostly foods) or activities (a walk in the woods or a game of chase), and how to carry out complex instructions ("take the can from the fridge and put it in the next room") (Dunbar, 1997, p.53).

On the other hand, as Barrow (1997, p.208) points out, the links between language and counting in traditional cultures have yet to be studied, so the jury is still out on the question of how far our propensities to count are "hard-wired" and how much they owe to cultural evolution.

PATTERNS

Another angle on the question of place of counting in qualitative analysis can be gained by reflecting on patterns. Researchers being introduced to the mysteries of qualitative analysis are often exhorted to search for and identify patterns in their data. Strauss and Corbin (1990, p.70) refer to patterns as groupings of several "dimensional profiles," that they, in turn, define in terms of how category's properties vary ("over their dimensional continuums"). Later they describe patterns as "repeated relationships between properties and dimensions of categories" (Strauss & Corbin, 1990, p.130).

In relation to some research on pregnancy and risk, Strauss and Corbin distinguish the properties and dimensions of the category "protective governing" as follows:

Property	*Dimensional range*	
Perceived risks	low	high
Pregnancy-illness course	on	off

They go on to offer an example of pattern formation, that shows how risk perceptions can vary with the course being followed in pregnancy, though introducing further "dimensions" to risk:

On-course lower-risk context
On-course, higher-risk context
Off-course, noncritical context
Off-course, critical context

Despite the complication of new dimensions, one can discern here a classic cross-tabulation of variables and values:

	Risk			
Course	Low	High	Noncritical	Critical
On course				
Off course				

The question is: how are these properties to be related (and a pattern established) if not through some assessment of the frequency with which the cells are filled in the table? Certainly numbers are not entirely irrelevant, for Glaser (1978, p.67) dismisses as "ungrounded" cells that have no data—arguing that "it becomes a mere logical elaboration of no worth to the theory to develop a type for such non-empirical cells," and adding that such cells are "of no theoretical use." This may seem a surprising claim, given that what is missing from our evidence may sometimes be more revealing than what is present. But it seems to follow from the insistence on inducing categories from data, rather than through logical deduction. Therefore grounding must involve (counting) the presence of at least some data in some cells. Moreover, before a pattern can be identified, we have to observe repetitions—that is, connections between categories that are observed in "case after case" (Strauss & Corbin, 1990, p.130). To record a repetition, you have to "count" it. So it seems that grounded theory may depend after all for its grounding on some number work, even if this tends to remain covert.

NUMBERS AND PATTERNS

The observation of patterns is part and parcel of theory generation, but patterns and numbers have a close affinity. Think of a knitting pattern, where the distribution of colors across a dress reflects the number of rows and stitches allocated to each color. The pattern of color in the dress may be regular, which is how we tend to think in terms of patterns in qualitative analysis. We can think of those patterns as repetitions. But patterns need not be regular. We can follow a pattern that produces an irregular shape in which a simple rule produces a quite complex and irregular outcome.

A topical example is the "Penrose tile"—topical because (as I write) it is currently the subject of a copyright suit over its apparently unacknowledged use in the design and production of toilet paper. The Penrose tile, in fact, consists of two diamond shaped tiles, one large and one small, and each is marked with a couple of black dots (see Fig. 6.5). The tiles can be joined up, but only in such a way that each corner with a dot touches only other corners with dots. This simple rule never results in a repeating pattern: the simple shapes (plus rule) produce infinitely complex results (Rucker, 1987, p.110), some of which have virtues (such as strength) that have attracted commercial interest (*Scientific American*, July, 1997).

Perhaps a better-known example is Conway's computer simulation of the "game of life." In this simulation, an initially random distribution of dark and light cells can evolve into a variety of patterns given simple transformation

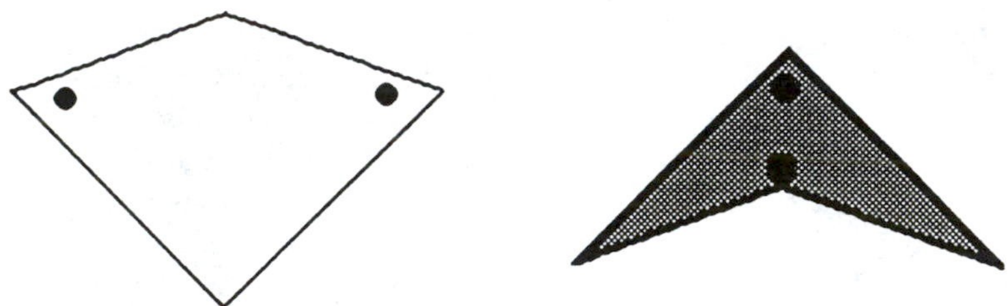

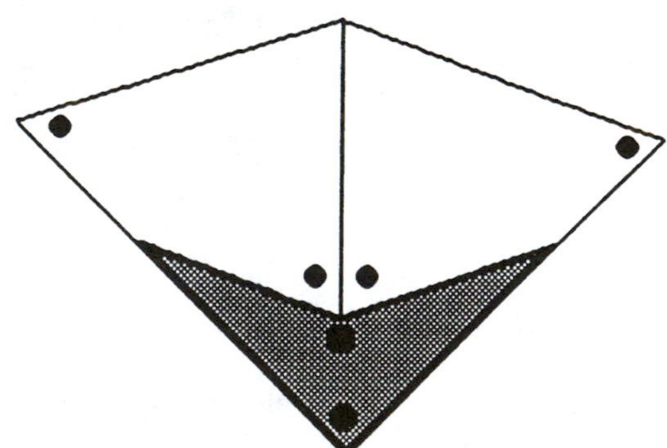

FIGURE 6.5 The Penrose tiles. Source: Derived from Rucker (1987, p.109).

rules that allow any vacant cell with three neighbors to become occupied (reproduction); any cell to become vacant that has less than two neighbors (death from isolation), or more than three neighbors (death from overcrowding). This process is known as a "cellular automaton" because, given definite rules, the computer automatically produces new patterns from an initial (random) distribution of cells. It suggests ways in which complexity can emerge and evolve into forms that "self-reproduce" over time and therefore mimic life itself. For example, some seashells seem to have some simple rules to determine the color of the next bit of shell they grow along their edge, while similar processes seem to govern plant growth. Because these patterns "evolve," it is not possible to ascertain in advance precisely what the outcome will be: "we can have simple starting conditions whose consequences we are unable to predict" (Rucker, 1987, p.313). Nevertheless we can grasp the pattern—that is, the simple rules which from an initial starting point produce an apparently complex outcome.

The idea of complexity here differs from its common interpretation as mere complication. We tend to think of the complexity of a system in terms of how complicated it is and therefore how difficult it is to describe. But complexity may also be understood in terms of the most concise account we can give of the rules that govern a system. The point is most easily illustrated with numbers. From a common sense standpoint, a string of numbers 1, 3, 5, 7, 11, 13, 17, 19, 23, 29, 31 may seem very complex. In fact it is generated by only one rule—just add the next number which is only divisible by itself and 1. Though the string may appear complicated, it is not complex because it can be accounted for in terms of a simple rule. Compare it with the following string of random numbers between 1 and 31: 26, 9, 21, 17, 27, 8, 20, 11, 8, 18. The only way we can account for these numbers is by giving a full description of the whole string. This string is more complicated but it is not complex as there are no rules to describe it. Complexity here is not about how complicated things appear on the surface but about how complex is the "pattern" which produces those complications. Complexity tends to be greatest where a system is "neither too orderly nor too disorderly" (Gell-Mann, 1995, p.59)—where we can identify some underlying rules which account for it, but these are not simple rules of the "add 1 to the previous number" variety. Thus complexity can be considered as a characteristic of the underlying "message" rather than the amount of surface "noise" in a system.

In social science, we still tend to talk of identifying patterns as though our evidence exhibited regularities, rather like the design of our carpets or wallpaper. We have yet to catch up with the Penrose-patterned texture of our toilet tissue. But this sense of pattern, as the underlying rules or instructions from which something is made, is perhaps the more promising in terms of theory generation. As Einstein put it, the point is not to describe the taste of the soup, but to explain how it has that taste. Instead of describing the distribution of colors of the dress, with whatever (ir-)regularities this reveals, we can look instead for the underlying set of instructions (rules) from which the garment is made. These may reduce apparently complicated and variable results to some simple rules and a given set of conditions. For example, suppose our instructions mirrored those discussed above, dictating when particular colors can be used according to rules like those of the Penrose tiles or the Conway game. We can understand (but not predict) the results if we can identify these rules together with the particular starting conditions, which may be arbitrary. In real life, of course, chance rarely deals a single hand, and other (random) factors may impinge on the evolution of an otherwise regulated process.

Searching for patterns in the data may therefore suggest two rather different processes, though both of them are susceptible to some sort of mathematical modeling. The search for what we could call surface regularities can be modeled through statistics, which provide a means of establishing frequencies (how often does x appear) and correlations (along with y). The search for underlying rules requires a different approach, as the surface relationships are complicated and irregular. Here, therefore, we have to turn from linear to nonlinear mathematics. When results of a process vary dramatically given small changes in initial conditions, we cannot explain them in terms of a uniform relationship between variables, such as the (linear) equation that governs the direction of a straight line in two-dimensional space. That can be modeled by a linear equation of the sort: $y = mx + c$, where the variables x and y are independent of one another. In linear equations, the outcome (the whole) can be predicted from the sum of the separate values (the parts). Sound and light act this way—so I can hear the phone above the sound of my typing and see the light from my lamp and the daylight from my window. The sound and light waves mix but retain their identities (Waldrop, 1994, p.64).

But what happens when variables do interfere with one another? For example, suppose you sell stock in anticipation of a market down-turn and so precipitate the very down-turn that you feared. The economy is non-linear in that many individual decisions (about jobs, spending, etc.) create a climate of boom or bust which in turn influences those decisions (Waldrop, 1994, p.65). In nonlinear systems, variables are not separate but react to each other through feedback, so that the outcome (the whole) cannot be calculated from the sum of the separate variables (the parts). This, of course, mirrors much of life—where so many apparently trivial actions can have so many enormous repercussions. Such interaction is a characteristic of open systems. Nonlinear systems do not yield neat lines on the graph paper—they produce dips and bumps. Kosko (1994, p.108) suggests that if you view a linear system as a flat sheet of paper, then you can see a nonlinear system as a crumpled sheet. Close to, the dips and bumps may even out and your local part of the crumpled sheet may look flat. But stand back and you realize your local part only looks flat because you ignore the whole picture.

Much of physical science has been concerned with understanding closed systems (first the earth, then solar system, and finally the universe) which scientists have tried to model through general laws. One result, according to Kosko (1994, p.108) is that we know "a great deal about linear math" but "almost nothing about non-linear math—except that almost all math is non-linear." However, one major constraint on nonlinear mathematics (the intricacies of calculation) has been reduced if not eliminated by the advent of the computer, with its enormous computational capacities (Waldrop, 1994, p.65). As a result, there has been burgeoning interest in analyzing complex dynamic processes (such as turbulence in fluids or phase transitions from one state to another) that previously eluded the mathematical competence (and hence the attention) of science. Such developments may well produce new mathematical tools for the social sciences (Eve, et al. 1997). Meantime, it is sufficient to note that counting may after all be relevant to the identifying patterns, whether considered as surface regularities or underlying rules.

The problem with counting, though, is that it seems to require the full coding that Glaser and Strauss are anxious to avoid. If counting can contribute to the identification of surface regularities or underlying rules, then partial coding may fail to take advantage of this potential tool. Take a simple example. The following diagram, which sets out the overall pattern of categorization across a range of cases—which might mean respondents, settings, or events, and so on (Fig. 6.6). The emerging pattern is based on an either/or judg-

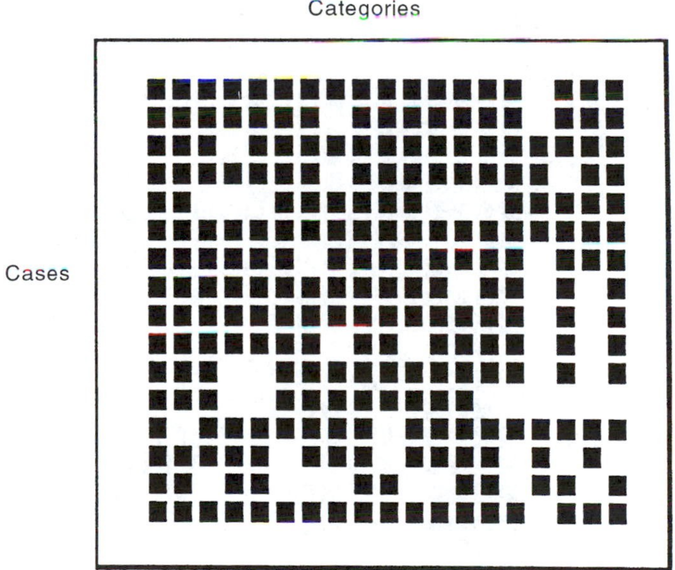

FIGURE 6.6 Categories assigned to cases (bivalent).

ment—has a category been assigned to a case or not? The pattern is informative, but the result is black and white—or "bivalent" (only two values allowed) in logical terms.

Now suppose we assign a number for each time a category has been assigned to a case (Fig. 6.7). For example, suppose we are comparing prenatal clinics and we have identified a range of categories (such as various levels of risk perception) that apply to the patients at the different clinics. Then we can identify the degree to which any clinic has patients who have been categorized in various ways. The white squares will tell us that no patients have been assigned the corresponding category. The gray squares will tells us that some patients have been assigned the corresponding categories. The darker the gray, the more patients have been assigned the corresponding categories. It seems plain that the pattern produced through the additional numerical information could be much more helpful in suggesting ideas about what is happening in the data.

The traditional objection to such forms of data reduction is that it obscures the meaning of the data in context. Here we have reduced all the categorized text segments to dots on a matrix. The advantage of so doing is that it may help to reveal underlying patterns in the data that may be obscured by the needless detail and complicating specificity of particular text segments. This is important if our aim is not merely to reproduce data, but to gain some

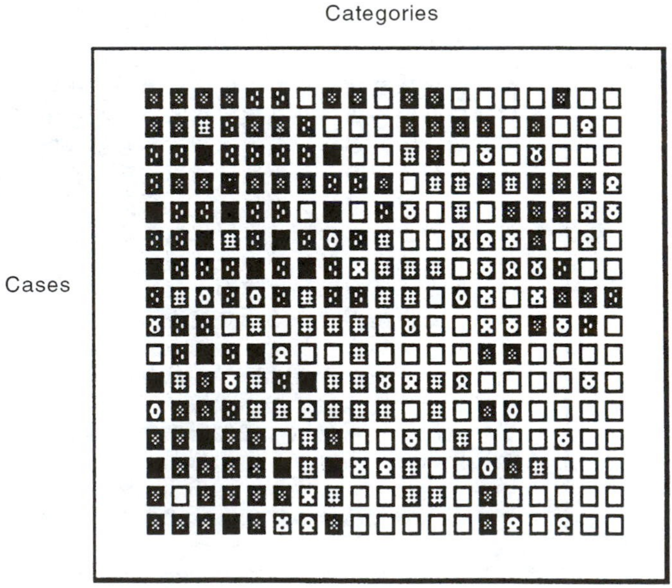

FIGURE 6.7 Categories assigned to cases (multivalent).

means of comprehending it. One of the heuristic values of the matrix presentation is that it may help to indicate where categorization has been concentrated. This may in turn assist in identifying the most fruitful lines of inquiry to pursue. The disadvantage of so doing lies in the distance between the dots in a matrix and the original data. However, the computer can allow us to reduce this distance by treating each dot in the matrix as a hypertext port of call through which we can access the original data in context.

These points suggest that a rigid commitment to partial categorization may mean missed opportunities. However, that in itself does not entail that full categorization is always relevant or required. So long as the aim is to generate theory, then it is the heuristic value of the method that matters. If counting (incidents, events, or whatever) helps—that is fine. But that does not mean it is essential. There may be some contexts in which it can contribute usefully and others where it would be a waste of time. This implies that a straight choice between partial and full categorization is not after all at issue. The question is rather where to draw the line somewhere between them—in what we could call a flexible form of categorization. Despite the emphasis in grounded theory on category saturation, there seems leeway for a more flexible approach along these lines, perhaps through developing a selective focus on particular categories.

IN CONCLUSION

Although coding is central to and even sometimes equated with analysis in grounded theory, it is difficult to clarify quite what this approach involves. On the one hand, we have the romantic vision that sees coding as an adventure, a game, a device for discovery. On the other hand, we have the mechanical and mundane tasks of systematic and painstaking analysis of evidence. One of the attractions of grounded theory is that at times we seem to be offered the pleasure without the pain. We can saturate categories without having to accumulate and analyze evidence systematically. We can identify repetitions in the data but discount its numerical dimensions. We can shift from open coding to selective coding in terms of a core category, shelving the onerous task of weighing alternative options against evidence. No wonder, then, that this heuristic of "systematic discovery" is so irresistibly attractive!

But despite this attractive heuristic gloss, questions arise nonetheless. How can we saturate categories without a rigorous and comprehensive confrontation with evidence? How can we identify patterns (and especially repetitions) without counting? How can we privilege one category without regard to the complexities of evidence and interpretation?

Subsequent debates, inspired by the widespread use of software tools to support coding, have wrestled with these issues, but without resolving them.

The debates over the different functions of coding—factual and referential, heuristic and representational, reductive and complicative—may lean toward a resolution in terms of the logic of discovery. But the authors cannot bring themselves to excise altogether the claims of evidence—even when these claims are cast within the framework of an alien paradigm. And those claims threaten to reintroduce the mundane and destroy the magical.

I have suggested a different resolution, that does not depend on contrasting the logic of discovery with the claims of validation. This resolution insists on the conceptual aspects of analysis, altogether denying the devil (mechanical coding) a place at the analytic table. The point is not to separate the mundane from the magical (or number from meaning) in order to preserve the mystique of the latter. Instead, we can marry the two by recognizing that the mundane and the numerical are always irreducibly conceptual: the point is to make this explicit. In qualitative analysis, coding (or categorizing, as I would prefer to call it) can never be mechanical. Even the most elementary tasks (like counting those ducks) require conceptualization. We cannot separate representational and heuristic functions. Nor should we discount the value of numbers within the logic of discovery. Even if we look for underlying patterns in terms of underlying rules rather than surface regularities, there is no reason to ignore the potential of mathematics in analyzing complexity. Those seeking to distance analysis from the mechanical and mundane aspects of coding are therefore attacking the wrong target. But perhaps they have created their own dilemmas, by adopting the term coding, with its inappropriate metaphorical and metonymic connotations, in the first place.

Freed from the framework of coding data, we can contemplate the various tasks of categorizing with more equanimity. We no longer have to apologize for categorizing our data systematically—for it need not mean we have thought any less about it. We can also contemplate other strategies, such as partial categorization, or even not categorizing at all. These strategies can be adopted or not, depending on our purposes. If we want to take full advantage of repetitions and patterns in the data, we may opt for full categorization, even if only to generate theory. But we may be content with a partial approach, looking only for evidence that suggests possible ways of conceptualizing the data. If we do want to ground our categories, we had best take account of all the evidence—though we may prefer to be flexible about where we concentrate our efforts. On the other hand, we may want to generate ideas rather than ground them, and if so we may not want to categorize at all. There are other methods of analysis, which may be more in tune with our purposes, such as a narrative approach.

If we do categorize, then we need to reflect more fully on how we create and use our categories. The distinctions between open, axiel, and selective (or core) coding appeal to a common sense view of the process of categorization. Begin by creating (and assigning) categories, continue by exploring connections between them, and conclude by focusing on an integrating core. These phases of inquiry naturally overlap to some extent—no one is advancing a proposal that rigid boundaries can be drawn between one phase and another. The problems I have noted above are more deep-rooted. They concern the way categories themselves are conceptualized as a method of analytic inquiry.

Two aspects of categorization discussed earlier are relevant here. One is the separation of open coding from axiel and selective coding—or, as I would prefer to put it, the process of creating and assigning categories from the processes of connecting and integrating them. Considered as a preliminary to connecting and integrating categories, open coding implies that categories can be created in isolation one from another. First we create them, then we connect them. This rationale also underpins the "code first think later" approach. It makes sense only within a correspondence model of classification. Once we acknowledge the limitations of that rule-bound approach and recognize the role of attribute clusters and underlying cognitive models in categorization, we cannot envisage categories as anything but connected. Therefore I suggested that rather than thinking in terms of category sets or categories in isolation, we think in terms of category strings.

The second aspect concerns the nature of these connections between categories. Here two points can be made. One is that we may need to consider those categories in which we have more confidence, namely those that are closest to a basic level. These may make a more useful point from which to start. At any rate, we should differentiate between categories in terms of how well-founded they are in prior experience. The second point is that we should recognize the value of holistic understandings rather than dismiss these as mere impressions or intuition. I suggested that holistic analysis can be rooted in the "rich and routine" understandings of our social world. For some, the very idea of an "holistic analysis" seems to be a contradiction in terms. But this is only so if we consider parts as separate from the wholes of which they are part. If, in contrast, analysis is considered as a way of analyzing parts as aspects of wholes, then we can reconcile holism with an analytic approach.

Nevertheless, the principle that data can be coded first (through open coding) and then codes can be connected and integrated afterwards (through axiel and selective coding) has had a powerful impact on the development and use of software for qualitative analysis. This methodological principle tends to legitimate the use of "code-and-retrieve" as a procedure which is preliminary to the theoretical tasks of conceptualizing the data. The language of coding (with its mechanical overtones) reinforces the idea that data can be managed and organized (through coding) prior to analysis, which then focuses on exploring the relations between categories. However, the problem that may then result lies in the construction of "an edifice of sophisticated reasoning" whose "weak link will always be the adequacy of the coding process" (Richards & Richards, 1995, p.456).

CHAPTER 7

Process and Causality

Unlike most research methodologies, grounded theory is about processes rather than people and places. According to Strauss and Corbin, "action and/or interaction lie at the heart of grounded theory" (1990, p.159), while "process" is "an essential feature of a grounded theory analysis" (1990, p.157). The analysis of process is seen as a distinctive feature of grounded theory, as indeed it often is of qualitative methods in general. In this chapter, I consider how "process" is analyzed in grounded theory with respect to the analysis of conditions and causes. In the next chapter, I consider how structural factors and human agency are incorporated into the analysis of interaction.

First, we have to consider the coding paradigm, which sets out the analysis of process in terms of conditions, interaction and consequences. We also have to consider the use of the conditional matrix and conditional paths in the analysis of process offered by Strauss and Corbin.

Struass and Corbin define process as "the linking of action/interactional sequences, as they evolve over time" (1990, p.157). This linking is accomplished by analyzing:

1. Change in conditions influencing interaction over time.
2. The interactional response to that change.

3. The consequences that result from that interactional response.
4. How consequences become conditions influencing the next. interactional sequence.

Here we have the coding paradigm in operation, with its framework of conditions, actions and consequences. Strauss and Corbin suggest that process can be expected to emerge "naturally" from the application of axiel coding; that is, from applying the coding paradigm. On the other hand, Strauss and Corbin also acknowledge that the perception and analysis of process requires something more than the application of a specific set of coding procedures:

> . . . unless the analyst is made keenly aware of the need to identify process, to build it into the analysis, it is often omitted or done in a very narrow or limited fashion.

This is because process "doesn't necessarily stand out as such in data" (Strauss & Corbin, 1990, p.143). Even though process "is very much there in the data, a part of any empirical reality," it remains elusive. One "feels its presence" in the data "even though one can't actually see it as such (Strauss & Corbin, 1990, p.144).

Strauss and Corbin also suggest that the analysis of process is not just "a simple description of phases or stages"—it requires something more (1990, p.147). But what? The immediate answer is that:

> It involves an in-depth examination of and incorporation of changed action/interaction into analysis, as this varies over time in response to changes in conditions (Strauss & Corbin, 1990, p.147).

This "in-depth" analysis involves capturing such aspects as:

- What leads to changes in conditions?
- How do responses vary as conditions change?
- What is the pace of change?
- What is the direction of change?
- How do variations in context (of time and place) affect the rate and degree of change?

Thus process is "the analyst's way of accounting for or explaining change." However, it would be "too overwhelming" to try "to capture all the change that occurs in reality":

> For us, as grounded theorists, it is a change in conditions of sufficient degree that it brings about a corresponding change in action/interactional strategies, which are carried out to maintain, obtain or achieve some desired end in relation to the phenomenon under study (Strauss & Corbin, 1990, pp.149–150).

The concept of change brings "time and movement into analysis." Time enters analysis in terms of duration, such as the lapse in time between changes in conditions (the stimulus) and changes in interaction (the re-

TABLE 7.1 Examples of Dimensional Ranges

Properties	Dimensional ranges	
Rate	Fast	Slow
Occurrence	Planned	Unplanned
Shape	Orderly	Random
	Progressive	Nonprogressive
Direction	Forward	Backward
	Upward	Downward
Scope	Wide	Narrow
Degree of impact	Great	Small
Ability to control	High	Low

Source: Strauss and Corbin (1990, p.150).

sponse). Movement enters the analysis in terms of the "passage" of a sequence of events (Strauss & Corbin, 1990, p.150). Change can also be analyzed in terms of various properties, each with a "dimensional range." Strauss and Corbin offer examples of dimensional ranges in Table 7.1.

Thus changes can be fast or slow, planned or unplanned, orderly or random, and so on. Presumably such variations can apply to either changes in conditions or changes in responses. For example, conditions may change rapidly as a result of some planning, while responses may evolve in a slower and disorderly manner.

Change can also be conceptualized in terms of "steps, phases, or stages." This involves identifying key "turning points" in the process. These turning points can be conceptualized as a "progressive" movement toward a particular result. As an example of progressive change, Strauss and Corbin cite their work on "coming back." This refers to the process of returning to a satisfactory way of life after a chronic illness, which they conceptualize as involving various stages "through which one must progress in order to reach come back" (Strauss & Corbin, 1990, pp.153–4). But according to Strauss and Corbin, change need not display such a progressive movement through phases. As an example, they cite the management of a chronic illness, which requires a process of constant adjustment to changing conditions where the whole aim is to keep the illness as stable as possible.

THE CONDITIONAL MATRIX AND TRACING CONDITIONAL PATHS

Hence process can be understood in terms of changing conditions and responses. But how can we analyze it? Strauss and Corbin emphasize the importance of interaction in relating conditions to consequences. This follows from their assertion that "the manner in which any phenomenon is expressed is through purposeful and related action/interactional sequences" (Strauss & Corbin, 1990, p.159). These interactional sequences are both embedded in sets of conditions and lead to specifiable consequences. Analysis involves identifying the conditions in which interaction is embedded through a conditional matrix, and then tracing the conditional paths through which these conditions result in specific consequences. The conditional matrix ranges from the most general international and national conditions right down to the group, interactional, and action levels (see Fig. 7.1).

Here interaction refers to processes of "negotiation, domination, teaching, discussion, debate, and self-reflection," while action (which can be strategic or routine) refers to

> . . . the active, expressive performance form of self and/or other interaction carried out to manage, respond to, and so forth, a phenomenon (Strauss & Corbin, 1990, p.164).

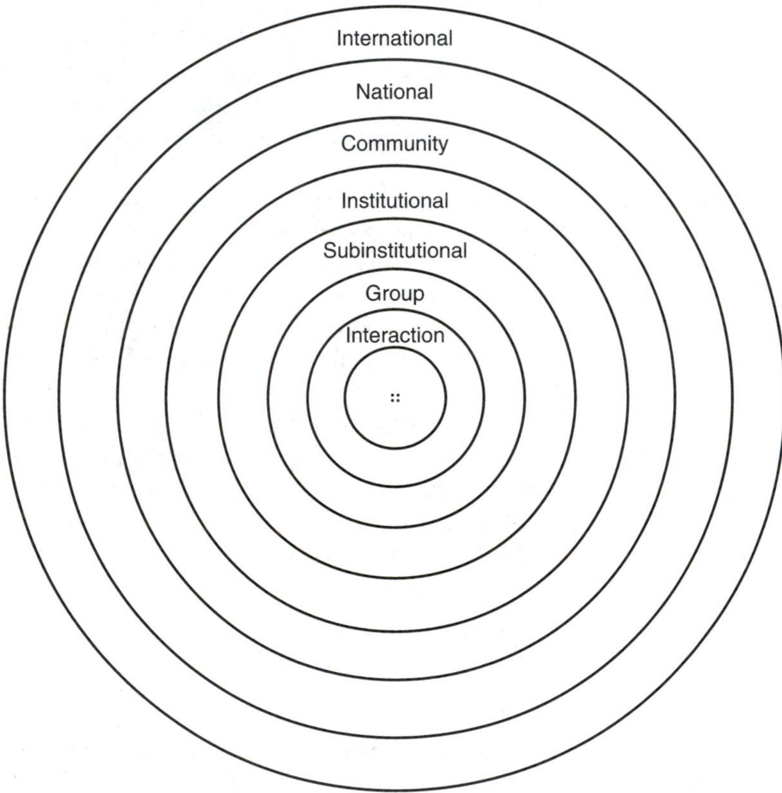

FIGURE 7.1 The conditional matrix (adapted from Strauss & Corbin, 1990, p.163).

Though the distinctions between conditions, action, and interaction may seem blurred, the main point is that conditions at all levels must be specified and linked. Nor is it enough to identify conditions in a general way—we have to connect specific conditions with consequences through their effect on interaction (Strauss & Corbin, 1990, p.167). This is called "tracing the conditional path," that is, relating interaction to specific conditions or specific conditions to interaction:

> To trace a conditional path, you begin with an event, incident, or happening, then attempt to determine why this occurred, what conditions were in operation, how the conditions manifest themselves, and with what consequences (Strauss & Corbin, 1990, p.168).

As an example of tracing a conditional path, Strauss and Corbin cite what could be called "the case of the missing gloves." This refers to an incident observed during a ward round in which the physician required (and insisted on obtaining) size 6 gloves to undertake an examination. The gloves proved unavailable. It turned out that gloves were in short supply in part because of the impact of the AIDS epidemic and newly published guidelines on national infection control. The gloves were therefore scarce and access to them was subject to careful monitoring. A pair was finally located in the recovery room.

Strauss and Corbin suggest that the conditional path in this case can therefore be traced from the immediate action (the doctor insists on size 6 gloves) through hospital and community to the national and international levels (as supply expands to meet the shortage in supply). In this case, the conditional path can show the specific connection between a change in general conditions (the AIDS epidemic) and managing the flow of work in a hospital ward. Of course, we could not hope to complete such a detailed analysis by tracing the conditional path of all conditions: "one would choose only those incidents that seemed especially pertinent to the central phenomenon under investigation" (Strauss & Corbin, 1990, p.171).

Where the requisite data are missing, Strauss and Corbin suggest that the researcher resort to "deductive thinking":

> . . . there may be times when the analyst is not able immediately to find evidence of process in the data. Either it's there, but not recognized as such; or there are insufficient data to bring it out. When this happens, the analyst can turn to deductive thinking and hypothesize possible potential situations of change, then go back to the data or field situation and look for evidence to support, refute, or modify that hypothesis (Strauss & Corbin, 1990, p.148).

However, they add later that "to be relevant, a condition or consequence has to be given meaning in terms of what you are studying (Strauss & Corbin, 1990, p.168). It is not enough to conceive of a potential condition (or consequence) in the abstract (that is, deductively): evidence of it must be finally found (and found repeatedly) in the data.

Glaser's Critique

Before we look more closely at this account of process, let us briefly consider what Glaser (1992) makes of it in his critique of the Strauss and Corbin version of grounded theory. One point, perhaps familiar by now, concerns the forcing of data to fit preconceptions. Strauss and Corbin identify a range of conditions, from global concerns to the most particular actions, and suggest that the links between all of these must be traced, at least for those incidents that are "especially pertinent." Also they identify aspects of change—such as differences in rate, direction, scope and so on—that ought to be included in the analysis. All this Glaser dismisses as "full conceptual description." He contrasts this approach with the focus in grounded theory on explaining (and only explaining) variations in the dependent variable. It is certainly striking that both the specification of the conditional matrix and the tracing of conditional paths are presented by Strauss and Corbin without any reference to the constant comparative method, which in grounded theory is intended to isolate those factors that might account for variation.

To Glaser, factors that fit a preconceived paradigm are simply not relevant (unless they emerge from the data) and can therefore be safely ignored. Nothing must be included in the analysis; or, as Glaser puts it, "there are no misses" in grounded theory (since there are no prior expectations). This applies even to process and change, for these may or may not emerge as significant in the course of a study. Glaser also objects to the conceptualization of change in terms of the coding paradigm which he suggests is no more than "the description of general motion in all studies . . . a general orientation that most of us are dedicated to" (Glaser, 1992, pp.90–1). Glaser is not objecting to the coding paradigm as such, which indeed (in the more general form of "the six Cs") he describes as the "bread and butter" theoretical code (of sociology) and the first to keep in mind when coding data (Glaser, 1978, p.74). Rather, he rejects the automatic privileging of this one coding family above others.

In his own work on theoretical codes, Glaser (1978, p.74) identifies process as a separate (if overlapping) coding family, comprised of stages, phases, transitions, trajectories, and so on. There he argues that "the analyst cannot talk of process and not have at least two stages in mind, for process is a way of grouping together two sequencing parts to a phenomenon" (Glaser, 1978, pp.74–5). Thus "the basic requirement to delineate a process is twofold: movement over time with at least two stages" (Glaser, 1992, p.90). Hence Glaser also objects to the suggestion by Strauss and Corbin that process can be conceptualized as without stages, or as being "non-progressive." It is precisely through the identification of stages (or phases) that it can contribute (but only if the data warrants it) to the emerging theory. Finally, he also objects to the argument that process in these terms is the way of explaining change, since this, too, makes a presumption about what has to emerge from the data.

Clearly there is some divergence between the original authors of grounded theory in their suggestions for analyzing process. On the one hand, we have Glaser's pluralistic approach to theoretical codes and his precise (but minimalist) account of process in terms of phases. On the other hand, we have the elevation of the coding paradigm to a key (or privileged) position, and a much more general account of process in terms of social change. About the only common ground shared by these positions seems to be a commitment to interaction as a key element in the analysis of process and change in grounded theory. Otherwise these conflicting accounts of process certainly raise more questions than they answer. At the very least, we are left to wonder whether the analysis of process is indispensable to grounded theory; whether process must be analyzed in terms of phases; whether this requires an analysis of process in terms of conditions, actions and consequences; and whether this demands specification of conditions in terms of all the levels of the proposed conditional matrix through the tracing of conditional paths.

Let us therefore consider first of all an area where there seems to be at least some common ground: that change can be analyzed in terms of conditions, actions, and consequences, as elaborated in either the coding paradigm of Strauss and Corbin or in "the six Cs" of Glaser's theoretical codes.

Let us consider the role of conditions and causes in this sequence. What are they and how can we identify them?

CONTEXTUAL, CAUSAL, AND INTERVENING CONDITIONS

Strauss and Corbin distinguish between causal, intervening and contextual conditions. In the course of tracing a conditional path, the analyst is encouraged to distinguish between those conditions that are causal, those which are contextual and those that refer to intervening conditions. Causal conditions are "those which lead to the occurrence or development of a phenomenon" (Strauss & Corbin, 1990, p.100). Taking the example of a broken leg, they note that pain may result from (be caused by) a broken leg, so that the broken leg can be seen as a causal condition of the pain. In this example, Strauss and Corbin seem to imply that causal conditions are observed directly and based on experience: we know for sure that a broken leg results in pain because we have all experienced how the body reacts to injuries of this sort. Note, though, that while it may seem obvious that a broken leg causes pain, the converse—that the pain is caused by a broken leg—may be harder to establish. If we experience pain, then (depending on circumstances) we may be less certain that it is the result in a break in the leg—this might require an X-ray to establish.

Causal conditions are also described as antecedent conditions, and can be identified through direct cues in the data (e.g., respondent uses "because") or by looking systematically for events preceding that under scrutiny.

Strauss and Corbin define "intervening conditions" as "the broad and general conditions bearing upon action/interactional strategies" or more generally as "the broader structural context" (1990, p.103). These intervening conditions can be facilitative or constraining and include such conditions as: "time, space, culture, economic status, technological status, career, history and individual biography" (Strauss & Corbin, 1990, p.103). With regard to the broken leg, for example, they suggest that spatial factors (broken while in the woods alone), temporal factors (so a long times passes), biographical factors (age, previous history of illness, philosophy of pain, etc.), and technological factors (available medication, etc.) could all make a difference.

Though these are the kind of conditions we might more commonly call "contextual," Strauss and Corbin use the term "context" in a rather curious way to refer to what they call "the specific properties of a phenomenon" (Strauss & Corbin, 1990, p.102). What does this mean? The formal definition of context that they propose does not seem to help much:

> Context: The specific set of properties that pertain to a phenomenon; that is, the locations of events or incidents pertaining to a phenomenon along a dimensional range. Context represents the particular set of conditions within which the action/interactional strategies are taken (Strauss & Corbin, 1990, p.96).

Here "context" is used in a rather idiosyncratic way to refer to the specific characteristics of an incident or event, which vary from setting to setting, as well as the surrounding events or conditions within which an incident occurs. Strauss and Corbin do allude to the more orthodox idea of context as "surrounding conditions," as is also evident in the following passage:

> Context, at the same time, is also the particular set of conditions within which the action/interaction strategies are taken to manage, handle, carry out, and respond to a specific phenomenon (Strauss & Corbin, 1990, p.101).

However, they again go on to equate the conditions of an event with the properties of a phenomenon. With regard to the painful broken leg, knowing the contextual conditions means:

> . . . we would want to know specifically when it was broken, how, the number of fractures, the types of fractures, whether sensation is present or absent from in the leg, and so forth. About the pain, we would want to know something about its trajectory, or course over time; also its duration, its specific location, its intensity and so forth (Strauss & Corbin, 1990, p.102).

We usually refer by "context" to the general conditions within which an incident or event takes place. We distinguish between the specific causes of an event and the more general conditions or circumstances that make that causal connection possible. For example, I cause the light to go on and off by operating the switch, but that is only possible because the light bulb is working, the switch is accessible, I am able to reach it, and a host of other conditions that make my action possible. Using context as Strauss and Corbin do, I might refer rather to the shade, intensity, and direction of the light itself. If the light is very bright, I might decide to dim it. In the common use of the term "context," the fact that I have a dimmer switch is part of the context which makes my act possible. In the Strauss and Corbin use of the term, the brightness of the light becomes the context in which I adopt my strategy of dimming the light. In another context (the light is dim) I might pursue a different strategy (and brighten it).

These different types of conditions in grounded theory can be summarized as follows (Strauss & Corbin, 1990, pp.100–103):

Different Types of Conditions in Grounded Theory

Causal	Those which lead to the occurrence or development of a phenomenon
Intervening	The broad and general conditions bearing upon action/interactional strategies
Contextual	The specific properties of a phenomenon

As conditions and causes are so central to the analysis of process, let us look more closely at the distinctions we can draw between different types of conditions.

Distinguishing Conditions

We are all familiar with the idea of imposing conditions on action. Take the warning: "if you are not back by midnight, you'll be grounded." This clearly imposes a condition ("be back by midnight") on action ("or you'll be grounded"). If the condition is fulfilled and we are back by midnight, we may be allowed out again. Everyday life is replete with such conditions, which we may create ourselves or which we may (more typically) recognize as existing or imposed constraints on our actions. If I want to write this book, I must first switch on my computer. If the publisher wants to sell the book, it must be marketed. If you want to read it, you must buy or borrow it. For particular actions to be possible, certain conditions must be fulfilled. The underlying logic of conditional analysis is an "if-then" statement: "*If* you are back by midnight, *then* you will not be grounded."

> Condition: *1. Stipulation, thing upon the fulfilment of which depends that of another; 2. (pl) Circumstances, especially those essential to a thing's existence.*

However, conditions can be a bit more complicated than it first appears. Let us stay with my first example: "if you are not back by midnight, then you will be grounded." This condition seems simple enough but, unfortunately, we can easily complicate it. Suppose the unhappy event occurred: last Thursday we missed the midnight deadline and we were grounded. Then can we explain being grounded in terms of failing to meet the condition? Yes—and no.

Yes—if unless the condition had been imposed, we would not have been grounded. But we cannot be sure that this would be so—perhaps we could have been grounded for some other reason—such as taking the car without permission or drinking and driving. There may in fact have been many other conditions (perhaps not made explicit) that we also had to meet in order to avoid being grounded, even had we been in on time. Thus meeting the stated condition (being in by midnight) did not guarantee that we would not be grounded.

Moreover, we could have been forgiven our sin and allowed out again, despite missing the midnight deadline. Suppose we only missed it by a minute or two. Just how late after midnight could we be before the punishment would be invoked? Could we have offered an excuse for being late, and what kind of excuse would have proved acceptable? Could we have lied in order to avoid the worst? And if we had lied, would we have been believed? What seemed a straightforward matter (back by midnight—or else!) becomes a little fuzzy at the edges. Now we have to take into account a range of other conditions as well—at the very least, we have to take account not just of the rule of being grounded if late, but also how that rule is interpreted and applied.

Suppose we ask whether this particular condition-consequence sequence of events was a reasonable one. This is an eminently reasonable question, given that there is always a potential conflict between the ruler (imposing a rule) and the ruled (from whom obedience is required). Suppose we challenged the rule as unreasonable, because no one else in the neighbourhood had to suffer such a restriction (this may not be true but in the likely absence of communication among parents it always makes a good argument). Then why were we grounded? Not because of a rule, but because of unreasonable parents! Or we could accept that it was because we obeyed the rule. We could argue (this is a favorite theme of Marxist scholars) that it was our own lack of resistance (or of ideological enlightenment), or our deferential relationship toward our parents and their rules, that explains why we were grounded.

Whether or not such rules are seen as reasonable or not may vary according to cultural context, family dynamics, and even the evolution of particular child-parent relationships. For example, a recent British study (Brannen *et al.*, 1994) found variations among families from different backgrounds in their fondness for setting boundaries and imposing discipline upon their children. Parents were also inclined to be more or less restrictive depending on the gender of the child. An earlier study (Backett, 1983) suggests that parents also take account of the individual history of their relationship with the child in making disciplinary decisions. For example, a normally obedient child may be more readily forgiven (and perhaps even encouraged to break rules a bit) whereas a normally disobedient child (seen as in need of restraint) may more likely be punished.

Although I have written of rules, this too proves rather presumptuous. Can we be sure that this condition-consequence link was governed by rule(s)? The "if-then" logic suggests as much—that if or whenever A happens (home past midnight), then B results (grounded). But life is not like that, because of the small matter of contingencies. Something usually crops up. Suppose Mum is the disciplinarian and because of her job she sometimes gets an unexpected call to leave town and she leaves Dad in charge. Dad just happens to be a more forgiving sort (or just lax) and does not stick to the rules. If Mum had been called out of town, we might not have been grounded. "If A then B" happens only "subject to other things being equal"—and life being what it is (a catalogue of contingencies) they rarely are. So any particular consequence can usually be attributed only to a specific configuration of conditions (that call did not come) and cannot be seen as an automatic consequence of the operation of general rules.

Were We Grounded Because We Came Home Late?

- *Other conditions may have had the same result.*
- *It may have depended on how the condition is interpreted and applied.*
- *Other conditions may have been more important.*
- *We need to take account of the cultural context, etc.*
- *We cannot ignore the specific configuration of events.*

It seems that a sequence of condition-consequence events summarized by an "if-then" rule may be rather misleading. We need to think a bit harder about the way conditions operate. Usually when we are asked to think a bit harder, we can resort instead to classification, and this case is no exception. For in light of the example above, we can make a number of useful distinctions about conditions.

First, we can distinguish between necessary and sufficient conditions.

Necessary conditions are those that must be present before an event can occur: if no condition, then no occurrence. For example, we could not be grounded for missing the midnight deadline if no one found out. So "finding out" is a necessary condition for being grounded. Sufficient conditions, on the other hand, are those that can produce a result all by themselves with no help needed from anyone or anything else. For a parent to find out that we missed the deadline, it may be sufficient that they heard us come in after midnight. And that is all that is required.

Conditions can be necessary, or sufficient, or necessary and sufficient—or they even may be unnecessary and insufficient. By and large, when an event is both a necessary and sufficient condition for something else to happen, we think of conditions as causes—but more of that later.

Conditions may be merely contingently related to a result and still act as conditions. Suppose the car broke down and that was why we were late. No one could plan for this contingency. But the car breaking down is neither a sufficient condition (we might have got home on time) nor a necessary condition (we might have been late anyway).

Necessary and Sufficient Conditions

- A condition is necessary if it must be present for a certain outcome to occur.
- A condition is sufficient if only it need be present for a certain outcome to occur.
- A condition is sufficient but not necessary if only it need be present for a certain outcome to occur, but the outcome may occur even if it is not present.
- A condition is necessary but not sufficient if it must be present for a certain outcome to occur, but other conditions must also be present.
- A condition is necessary and sufficient if it must be present for a certain outcome to occur, and only it need be present for that outcome to occur.

As we have seen, Glaser and Corbin distinguish between causal and contextual conditions. Insofar as the rule that we are "grounded if late" produces the anticipated effect, it might be described as a causal condition. But as we saw, other conditions may have contributed, such as having an unreasonable parent or having been brought up in a restrictive culture. These conditions may be regarded as setting the context in which the rule "grounded if late" is formulated and applied. The context may range from more immediate conditions (Mum was in a bad temper that night) through middle range (the exams were approaching) to much more general conditions (our parents are very strict because of their religion). In a sense, contextual conditions are limitless—for example, we could say that the ability to communicate through language, or precise methods of keeping time, are conditions for the "grounded if

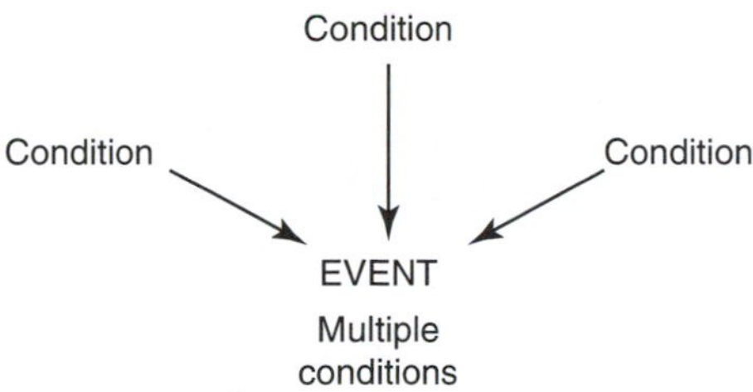

FIGURE 7.2 Multiple conditions.

late" rule. So an important question, if contextual conditions are potentially unlimited, is where and how we can draw the line.

There may be occasions where a single condition stands out as causal—for example, if the only time we were ever grounded was when our parents had just had a row. But even then, we may doubt whether one condition could ever be regarded as the sole condition of a specific effect, as contextual factors may come into play. In general, a range of conditions combine to have more or less impact on how events unfold. Sometimes any one of a number of conditions could produce the same result. So there are multiple ways of arriving at that result (Fig. 7.2).

On other occasions, the result arises only if a number of conditions are met. A particular combination of conditions—or conjuncture—is required to produce the result (Fig.7.3).

Thus we can distinguish between events that we tend to attribute (rightly or wrongly) to a single condition, from those we account for in terms of multiple or conjunctural conditions.

Once we are alerted to multiple conditions, we may readily find conditions intervening between the condition we start with and the consequence that results. In general, the need to apply rules or implement policies constitute intervening conditions that explain why results seldom match our expectations. Things can literally "get in the way."

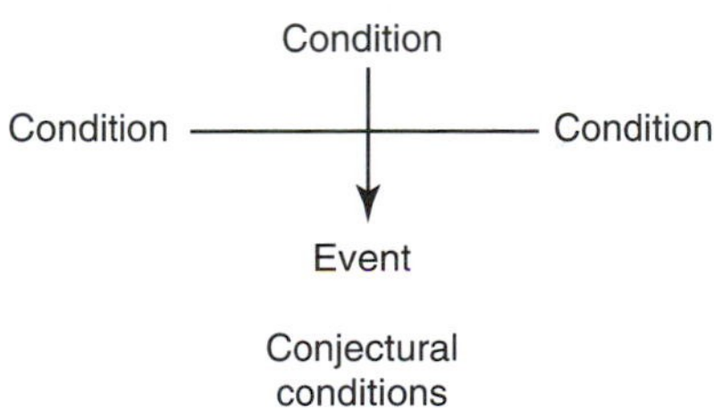

FIGURE 7.3 Conjunctural conditions.

Different Types of Conditions

- *A contextual condition can contribute to the occurrence of an effect but cannot itself produce that effect.*
- *A causal condition can produce an effect.*
- *A single condition is necessary and sufficient to produce an effect.*
- *Multiple conditions can each independently produce an effect.*
- *Conjunctural conditions must combine to produce an effect.*

Another problem posed by multiple conditionality is that of confounding conditions (though it is more usual to talk of confounding causes). The problem here is that we may mistake one event as the condition of another event when in fact it is not. There may be other conditions which explain why the two events seem related. For example, suppose all daughters are grounded for being late, while sons are always let off. May we draw the conclusion that gender is a condition for being grounded? This seems reasonable, but suppose we find out that all the daughters are taking exams and none of the sons are. Then we may reasonably ask whether the condition for being grounded is not gender but taking examinations.

Another distinction we can draw is between local and general conditions. The local conditions relate to a particular result—a specific action or event that happens at a given time and place. Why were we grounded last Thursday? That was (and could only be) the result of a combination (or configuration) of conditions unique to that particular event. Note that it is the configuration of conditions that is unique; the conditions themselves need not be. Once we characterize an event in terms of time and place (that is, at a particular conjuncture) both the event—and the configuration of conditions that precipitates it at that time and place—must be unique. We can safely assume that this is so, even if different observers might record the event as happening in different times and places—unless and until time travel becomes a possibility.

We can explain a particular event in terms of a local configuration of conditions (Fig. 7.4), but this may not always be the most useful way of trying to figure out what has happened.

At other times, we may want to consider the general conditions that account for a particular instances. Rather than referring to a specific configuration of conditions, we prefer an account in terms of more general conditions (Fig. 7.5)—"that's what happens when I'm caught coming in late." We may sometimes opt for an account in terms of local conditions and sometimes for an account in terms of general conditions, depending on what kind of explanation we find most useful.

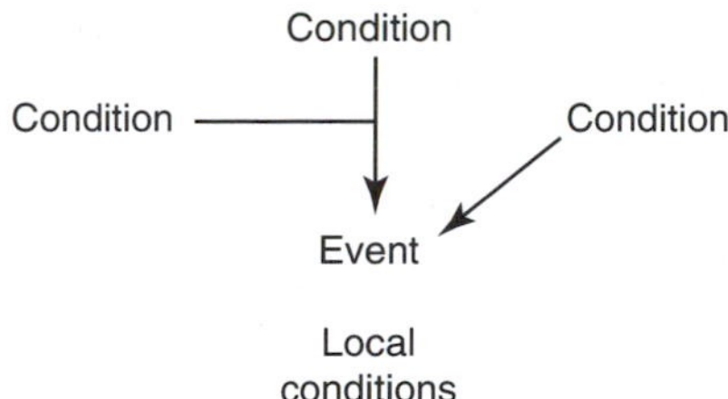

FIGURE 7.4 Explaining an event through local conditions.

If we can explain a particular event in terms of either local or general conditions, what about explaining more than one event? Perhaps all our friends were grounded last night as well. Again, we can account for these events in terms of either local or general conditions. This is because a local configuration of conditions can produce multiple consequences. For example, there may have been a party on Thursday night that went on too long, so everyone got back late. Or this may be exam year, which is why our parents (and others) are being so strict. These are local conditions that can account for a number of events. On the other hand, we may be more inclined to account for these events in terms of general conditions—such as the conflict between the claims of youth culture and educational aspirations, or a gap between the generations.

This raises the interesting question of when accounts in terms of local conditions might be more useful than accounts in terms of general conditions—and vice versa. Social scientists sometimes argue that they can only account for events one way or the other—though they have yet to agree on which! It might seem more productive to think through the appropriateness of different kinds of explanations for particular purposes than to try to "legislate" some as legitimate and others as beyond the pale. Sometimes we may want to focus on the event itself and at other times on the event as an example of something else, of something more general. Both of these seem reasonable ways of trying to account for events—but this is an issue I return to in the next chapter.

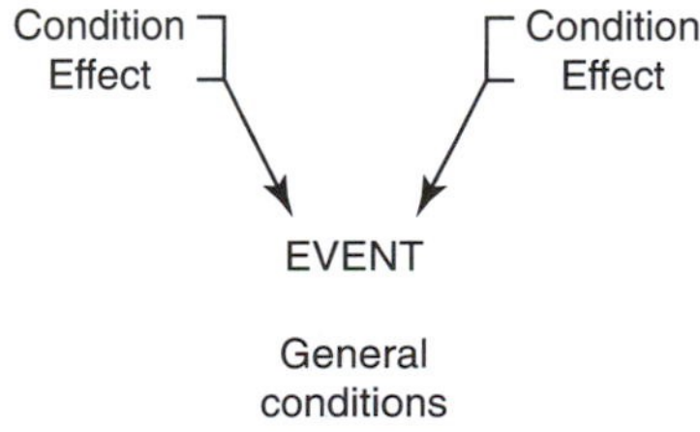

FIGURE 7.5 Explaining an event through general conditions.

Meantime, having examined different types of condition in some detail, we can now turn to the question of how to identify causes.

Causal Conditions

As we have seen, Glaser and Strauss refer to causal conditions rather than causes, and these are seen as one type of condition (in contrast to intervening and contextual conditions). You may recall that intervening conditions referred to "the broad and general conditions bearing upon action/interactional strategies" or, more generally, as "the broader structural context" (1990, p.103) which may facilitate or constrain interaction. As for contextual conditions, these referred to "the specific properties of a phenomenon" (Strauss & Corbin, 1990, p.102) that affect the way interaction unfolds. Recalling the example of the broken leg, where and when it occurred was described as an intervening condition, how the leg was fractured was described as a contextual condition, and that broken leg was painful was described as a casual condition.

These distinctions do not seem consistent with our usual practice, and if anything, they reverse it. We would usually refer to conditions of time and place as contextual, and the nature of the fracture as an intervening condition. How do we distinguish the two? The term "intervening" gives a clue, as it implies that these conditions come between cause and consequence. In other words, an intervening condition cannot precede or be coincident with a causal condition in a temporal sequence, but must follow it. On the other hand, we think of contextual conditions as "setting the scene" for a (causal) sequence of events and therefore contextual conditions must precede that sequence or at least be coincident with it. Conditions that follow a sequence may be described as contextual, but only in the sense that they may affect our understanding of events, and not the events themselves.

Temporal Sequence

This suggests that our ideas of causality are bound up with conceptions of time. We assume that consequences follow causes but do not precede them. As Strauss and Corbin suggest, causes are antecedent: "processes which, once started, end up producing a particular outcome at a later point in time" (Marsh, 1988, p.224).

Certainly, we have difficulty with the idea that one event can follow another in time and yet still be the cause of that event. This is true even where an outcome is anticipated and the initial action can only be explained in terms of the motivation to achieve that outcome. This does not really overthrow our ideas of temporal sequence, because there still has to be someone who anticipates, and that anticipation still has to precede the event which is anticipated. You would (or should) dismiss out of hand someone who said they acted in anticipation of an outcome, if they could only have had relevant knowledge of that outcome after the event. The intention has to precede and anticipate the action.

I say "should," because in practice we all make mistakes of this sort. Retrospective reasoning is the stuff of everyday life because it can provide comfortable post hoc rationalizations for the actions we have undertaken. Thus politicians are never slow to claim credit for the favorable but unintended results of policies designed for another purpose. In this respect, they display in public the same disposition for congenial rationalization to which we are all inclined in private. We are often disposed to disown responsibility for less than desirable consequences.

We can get away with this kind of thinking because our faith in causal sequence—that causes must precede consequences—tends to be stronger where we can offer a clear causal account. When causal explanations are not available, then it seems that teleological ones will suffice—so long as we have an explanation of sorts. Unfortunately, our demand for explanations (far) exceeds our capacity to supply them. And so astrology flourishes, apparently appealing to presidents as well as to paupers—in the hope that outcomes can be known (and not just anticipated) in advance. Even physicists are not immune to this kind of thinking—witness recent "explanations" of the universe in terms of anthropic principles concerning the requirements for the evolution of (intelligent) life (Gribbin, 1993, pp.185–6), suggesting that some 15 billion years ago, the universe has somehow been designed for our benefit (cf. Darling, 1989). The problem with this kind of explanation is that it really explains nothing, since not only temporal sequence is required to establish causality, but also some sort of intelligible process that links cause to effect. In this respect, astrological appeals to "fate" or theological appeals to "divine creation" are unintelligible since they are intrinsically mysterious and we cannot even begin to explain how they might work.

Thus temporal sequence is important, but it is only a necessary rather than sufficient condition in establishing causality. Clearly knowing A precedes B does not in itself establish A as a cause of B. So we still have to ask how causal conditions are identified in grounded theory?

GROUNDED THEORY AND ANALYTIC INDUCTION

Glaser and Strauss contrast their approach to causality in grounded theory with the methods of analytic induction, which they describe as:

> . . . concerned with generating and proving an integrated, limited, precise, universally applicable theory of causes accounting for a specific behavior (e.g. drug addiction, embezzlement) (Glaser & Strauss, 1967, p.104).

Analytic induction, in other words, aims to produce general causal explanations of behavior. According to Glaser and Strauss (1967, p.168) it does this by testing hypotheses against data for carefully selected cases, then reformulating hypotheses (or redefining the phenomena to be explained) "by constantly confronting the theory with negative cases" until a relationship is established that holds for all cases. Underpinning analytic induction is the method of agreement, as putative causes can be discounted if a "negative instance" is found. One negative instance—for example, where the putative cause is absent even though the effect is present—is enough to disconfirm the hypothesis.

The aim of analytic induction is to produce scientific generalizations, that consist of "conditions which are always present when the phenomenon is present but which are never present when the phenomenon is absent" (Cressey, 1953 cited in Hammersley, 1989, p.169). However, if it proceeds by observing effects and then checking whether a putative cause is present, analytic induction only addresses the first side of this equation. It cannot confirm sufficient conditions but only necessary ones. To test for whether a condition is also sufficient to produce an effect, we also need to check for cases where the condition occurs and check whether the effect is present (Hammersley, 1989, p.195).

At this point, a short detour through the classic methods of establishing causality through agreement and difference may prove useful.

The Method of Agreement

There are various procedures through which causality can be inferred from concurrence. One is Mill's "method of agreement," which Ragin (1987) explains as follows:

> . . . the method of agreement argues that if two or more instances of a phenomenon under investigation have only one of several possible causal circumstances in common, then the circumstance in which all the instances agree is the cause of the phenomenon of interest. . . (Ragin, 1987, p.36).

In other words, take an event which always precedes the phenomenon as the cause of it:

> . . . if an investigator wants to know the cause of a certain phenomenon, he or she should first identify instances of the phenomenon and then attempt to determine which circumstance invariably precedes its appearance. The circumstance that satisfies this requirement is the cause (Ragin, 1987, p.36).

Ragin suggests that it does not really matter whether we speak here of a single cause or of a specific configuration of causes, so long as either the cause or the configuration of causes fulfills the requirement that it "invariably precedes" the consequence.

One way to identify which cause or configuration of causes "invariably precedes the consequence" is to eliminate the alternatives. Suppose we have rival explanations of why students have been grounded—such as their gender, their academic commitments, or restrictive parental attitudes. Then we can eliminate any of these possible causes if we find students have been grounded even where this factor is absent. For example, if we find male students are grounded as well as female, this bodes ill for a causal explanation based on gender. If we find students are grounded regardless of their academic commitments, then we can eliminate this as a factor too. If we find students are grounded even where parents are not judged to have restrictive attitudes, then this too must go. The only factors that can figure as causes are those that survive this process of elimination. In this case, none do. However, suppose we had found that those who were grounded were invariably female with academic commitments and restrictive parents. Then we might infer that this configuration of causes produces the effect.

One problem with this kind of inference is that it may be based on a spurious relationship between the putative cause(s) and consequence. Suppose we always find where students are grounded that academic commitments are strong. Both strong academic commitments and being grounded may be effects of restrictive parenting. We can never be sure whether some other causal condition may be at work, which explains why both putative cause and consequence concur. Or to take a more clear-cut example, suppose that whenever I change the time of one clock in the house, I change the time of all the clocks. If I have three clocks and one does not keep time accurately, then using the method of agreement we could eliminate the clock that keeps inaccurate time, but the other two will concur. Whenever we observe the time on one clock, we can observe the same time on the other. As the method of agreement is based on identifying such "patterns of invariance" (Ragin, 1987, p.37), can we claim that one is the cause of the other? No, because the concurrence is a product of another factor—my setting the time on these reliable clocks.

Another problem arises from the ubiquity of multiple causation (that is, various causes can produce the same effect) because then the method of agreement can produce misleading results. Suppose being grounded is sometimes a result of restrictive parenting and at other times a result of strong academic commitment. Suppose also that there are families where academic commitment is strong, but parents are not restrictive and vice versa. The method of agreement requires a constant conjunction of cause and effect, but sometimes students may be grounded although academic commitments are weak (because parents are restrictive), and sometimes students may be grounded although parents are not restrictive (because academic commitments are strong). So these causes would both be discounted by the method of agreement (cf. Ragin, 1987, pp.37–8).

The Method of Difference

The method of agreement can be strengthened by the method of difference. This goes beyond the identification of concurrence to examine whether an effect is absent when the putative cause is also absent. The reasoning is that if a cause is present, its effect should also be present. If we found a constant concurrence between being grounded and restrictive parenting, we would be more confident in restrictive parenting as a cause if we found out that students are not grounded when parents are not restrictive. In Table 7.2 all our cases should fall into A and D. First we check using the method of agreement whether in all cases where parents are restrictive students are grounded (all cases are in A rather than B). Then using the method of difference, we check whether in all cases where parents are nonrestrictive students are not grounded (all cases are in D rather than C).

We can summarize this procedure in two "if-then" rules. To establish restrictive parenting as a cause, we must establish that:

- If parents are restrictive then students are grounded (using the method of agreement)
- If parents are not restrictive then students are not grounded (using the method of difference).

Our confidence in restrictive parenting as a cause will be increased if we can also eliminate some rival explanations. For example, suppose strong academic commitment is a competing factor, for we find that in cases where parents are restrictive and students are grounded (in A), academic commitment

TABLE 7.2a Analysis by the Methods of Agreement and Difference—Example

	Restrictive	Non-restrictive
Grounded	A	C
Not grounded	B	D

TABLE 7.2b Analysis by the Methods of Agreement and Difference—General

	Condition present	Condition absent
Effect present	A	C
Effect absent	B	D

is strong. If we find evidence of strong academic commitments among those cases where restrictive parenting is absent and students are not grounded (in D), then we may reject strong academic commitments as a factor.

Unfortunately the method of difference, while giving further grounds for causal inference where we are dealing with single causes, still cannot cope with multiple or conjunctural causation. If either restrictive parenting or strong academic commitment can act as causal conditions, then we may find cases where parents are not restrictive but students are grounded (that is, in C). If it is a combination of restrictive parenting and strong academic commitment that counts, then we may find cases where parents are restrictive but students are not grounded (that is, in B). It is only where a single cause produces an effect that we can expect all cases to fall into A or C.

So inference of causation from concurrence remains a problem, despite the methods of agreement and difference. There are no cast iron grounds on which to make such causal inferences. The problems are especially severe where (typically) causes are multiple (many different factors can produce the same effect) or conjunctural (several causes combine to produce the same effect).

These difficulties in analyzing multiple and conjunctural causation may explain why Glaser and Strauss seem wary of hitching their wagon to the classic methods of analyzing causation. Despite the emphasis on establishing patterns of similarity and difference through constant comparison, Glaser and Strauss discount the possibility or, indeed, the desirability of trying to establish single causes through grounded theory. Glaser and Strauss argue that, in contrast to analytic induction, grounded theory is not concerned to establish causal generalizations per se.

This is, first of all, because grounded theory is not concerned to establish (that is, test) theoretical propositions: thus "no attempt is made by the constant comparative method to ascertain either the universality or the proof of suggested causes" (Glaser & Strauss, 1967, p.104). We can think of the "universality" of a cause in terms of its necessity—that a condition is not only sufficient but also necessary to produce a given effect. This requires evidence that a cause is invariably present whenever an effect is present. In grounded theory, however, if a cause is missing, it is simply explained (or explained away) as a product of uncontrolled variation. The emphasis is on identifying all the factors that may produce a particular result—in other words, on exploring multiple causation. Thus grounded theory is concerned only with the identification of sufficient and not necessary conditions (for a slightly different argument arriving at the same conclusion, see Hammersley, 1989, p.202).

Second, Glaser and Strauss reject causal generalization as a goal of grounded theory because it is has a wider concern with analyzing what they (rather vaguely) term "general problems" rather than an exclusive preoccupation with causality. This suggestion perhaps hints that grounded theory is more concerned with analyzing complex multicausal processes as a whole (that is, with conjunctural causation) than with identifying and isolating particular causal factors. In examining relationships among variables in quantitative analysis, Glaser and Strauss argue that in grounded theory attention should focus instead on the joint effects of multiple conditions:

> . . . "holding constant" is a notion used in verification of theory, when the analyst is trying to reduce the contaminating effects of any strategic variable not in focus with his variable of interest. To view [data] in terms of joint effects of two conditions on a third lends itself better to generating theory, since no variable is assumed a constant; all are actively analyzed as part of what is going on (Glaser & Strauss, 1967, p.218).

The "constant" in the "constant comparative method" does not refer, then, to efforts to hold constant conditions that might contaminate the analysis (for example, by producing spurious correlations). Nor can it refer to constant concurrence between categories (or variables), since the "equivalence of indicators" means these need not be defined consistently across a range of cases.

Not content with cutting constant conjunction out of the analytic process, Glaser and Strauss also negate the notion of negative instances. Whereas in analytic induction (and theory testing more generally) negative instances are usually considered as grounds for rejecting a hypothesis—or at least as something to be explained away as a deviant case—in grounded theory they are welcomed as a source of theoretical variation and diversity. If previously established relationships are not found in the cases produced through theoretical sampling, this does not mean that these relationships do not hold, Glaser and Strauss argue, but merely suggests some new conditions under which variation occurs.

Thus grounded theory does not support analysis in terms of the method of agreement, since neither positive instances (that is, those in A of Table 5.1) nor negative instances (those in B) can be used to identify causal relationships. The positive instances are contaminated by the lack of consistency in categories across cases, while no single causal factor is identified that could be rejected through the identification of negative instances.

Does grounded theory support analysis in terms of the method of difference? The method of difference depends on a theoretical specification of relevant cases, where neither the putative cause nor the putative effect are evident:

> The examination of negative cases presupposes a theory allowing the investigator to identify the set of observations that embraces possible instances of the phenomenon of interest (Ragin, 1987, p.41).

Theoretical sampling seems to proceed along just these lines, as it allows the researcher to identify possible areas of inquiry (such as hospital wards) without any presumption about the nature of the relationships that may hold. However, theoretical sampling is not orientated to identifying negative cases:

> The sociologist does not merely look for negative cases bearing on a category (as do others who generate theory); he searches for maximum differences among comparative groups in order to compare them on the basis of as many relevant diversities and similarities in the data as he can find (Glaser & Strauss, 1967, p.56).

This sampling process is presented in two stages, first including groups with minimal differences (to establish "the basic properties of a category") and then those with maximal differences (to bring out the widest possible variation in categories). The aim, though, is not to use differences to negate established relationships but rather to recognize and incorporate new conditions. Thus the observation of family involvement in caring for dying patients in a different setting (a Malaya hospital) is simply incorporated into the analysis as an additional variable to be considered (Glaser & Strauss, 1967, p.57).

Ragin (1987, p.42) suggests that the use of contradictory evidence to elaborate rather than amend relationships is typical of case-oriented methods—even those which start out in search of invariant relationships, since these tend to turn out either "circular or trivial." Rather than rejecting initial hypotheses, researchers using case-oriented methods tend to look for ways "to refine their arguments and try to effect a better fit with the evidence" (Ragin, 1987, p.43). They can do so by reformulating hypotheses (as in analytic induction) to take account of additional conditions that explain variation from the relationships initially hypothesised. Or they can redefine the phenomenon under study (again as in analytic induction), for example, by redefining the putative effect more narrowly or by differentiating the effect into different types, whose relationships with various causal factors can then be examined. In such cases, the methods of agreement and difference are used only as "rough guidelines for the conduct of comparative inquiry" to generate a dialogue with evidence that in turn stimulates the development of new theories (Ragin, 1987, p.44).

Clearly, grounded theory falls into this general pattern of case-oriented research, though with some differences. The logic of causal analysis through the methods of agreement and difference is much attenuated in grounded theory, with apparently little effort to control conditions (for example, through case selection) so that analysis can focus on a very limited range of factors without worrying about the possible impact of other variables. The identification of causal connections through constant conjunction is weakened by the emphasis on heterogeneity and lack of concern for consistency in the use of categories across cases. And negative cases are not used to reject relationships, but only to qualify them.

In general, these departures from the logic of comparative analysis are justified by their virtues in generating theory that is conceptually both dense and diverse. However, these departures do raise questions about the efficacy of causal analysis in grounded theory. First, the logic of comparative analysis apparently continues to underpin the process of generating theory, for Glaser and Strauss continue to rely on comparison in terms of similarities and differences within and between categories across cases. However, given this methodology, confidence in the identification of causal relationships (setting verification to one side) may be undermined by the lack of consistency in and control over conditions. Just how well grounded can the generated theory be in these circumstances?

Moreover, there are few references to what alternative methods of analysis might be undertaken, notably of multiple and conjunctural causes. It is not clear how these more complex forms of causation can be dealt with effectively through the method of constant comparison, whose logic remains attuned to the specification of single causes. How can comparison allow us to disentangle such complex multiple and conjunctural conditions from partially correlated but otherwise wholly unconnected events?

The problem is rendered more serious by any claim that grounded theory can verify as well as generate theory. Causal relationships that can be no more than hypotheses—and perhaps not as firmly grounded hypotheses at that—may be presented as fitting the available evidence and in need of no further examination. Although inference based on a rigorous application of the methods of agreement and difference may be inadequate, particularly for multiple and conjunctural causation, without that rigor (or without some alternative methodology) we are left to wonder how any causal conditions in grounded theory can be identified or established with any confidence.

On the other hand, the analysis of similarity and difference in grounded theory is allied to an experiential approach that is rooted in practical knowledge of the social world under study. Let us consider whether this more practical "know-how" acquired through direct experience may provide a more secure foundation for the analysis of causality and process. As we have seen, Glaser and Strauss place a great deal of weight on the direct observation of interaction and how it changes over time. The weakness of causal inference from similarities and differences observed through constant comparison may be offset by the strength of direct observation of causal efficacy over time. Although Glaser and Strauss no more than allude to this approach, it is worth exploring in some detail as an alternative basis for grounding the analysis of causal process in grounded theory.

Intelligibility

To most observers, the causal world we inhabit seems anything but mysterious—much to the disappointment, perhaps, of mystics and magicians (not to mention some philosophers and social theorists). This is because causality is informed by intelligibility: we can understand very well how a cause produces an effect. This is as true of our everyday, common sense world as it is of the world of science. That does not mean, of course, that common sense knowledge of how things work has to emulate scientific knowledge. We do not need to grasp how electricity works to understand why the light comes on when we flick the switch. But we do need some sense of how the light and switch are connected—via electricity—so that flicking the one produces such a desirable result in the other. Similarly, we do not need to know how a petrol engine works, but we do need to have some sense of what drives a car (the engine) or what makes it turn (the steering wheel). Without this working knowledge, the sequence of events might seem truly mysterious (witness the cargo cults). With this knowledge, causality seems mundane. Even if we personally have only the vaguest idea of how things work, we may know someone somewhere has the requisite knowledge to produce and repair them. Hence our ease (and ignorance) in a technologically sophisticated world of televisions, lights, compact disks, and cars.

Lakoff suggests that our first intimations of causality come not from technical knowledge, but from bodily experience. The visual world appears when we open our eyes and disappears again when we close them (though the tactile world is still there—a point some philosophers are inclined to overlook). We learn that the push and pull of muscles can produce certain increasingly sophisticated effects—we can roll over, sit up, crawl, totter, walk, and (all too soon in the case of my own children) climb. We learn to vocalize sounds and the reactions these can produce. We learn how to satisfy hunger and thirst. We learn how to pick things up—and drop them. The intelligibility of these causal sequences does not have to be explained so much as experienced—because we ourselves initiate the action and experience the consequences. Thus it is experience (and not inductive inference) that first renders causality intelligible.

If the link between events is not intelligible, it becomes difficult to distinguish causality from mere correlation. As everyone knows (but many conveniently forget) events may be correlated without being causally connected, as the example of two clocks synchronized to keep time illustrates (Marsh, 1988, p.224). We can predict the time on one clock from the time on the other, but altering the time on one has no effect on the other. The regular concurrence of two events, like the clocks keeping time, need not imply any causal link between them. Therefore we cannot (logically) infer a causal link merely from the regular concurrence of events. Yet that is precisely how many social scientists try to proceed.

Our confidence in causal inference requires an intelligible link to be established between cause and effect. So far we have considered intelligibility only in terms of the logic of comparative analysis—using the (non-)concurrence of events as a basis for inferring relations of cause and effect. However, this is not the only (and perhaps not even the principal) way in which the intelligibility of causal relations can be established. Indeed, given the inability of the methods of agreement and difference to cope with causal complexity, it seems that an approach that relies so heavily on establishing patterns of similarity and difference is likely to prove unproductive. Are there other ways of identifying or establishing causality to which we can turn?

Fortunately, the intelligibility of a causal relationship may have very little (if anything) to do with the regularity with which events concur. We may not need to rely on an invariant pattern in which cause precedes effect. For one thing, we tend to think about causality in terms of powers rather than patterns. As suggested above, our first intimations of causality probably stem from the exercise of our own emerging powers to move our bodies, manipulate our environment, and make demands on others. That bodily experience provides an early but persistent insight into causal processes—we can understand directly how things work. As our prowess increases, intelligibility requires more abstract and acquired forms of knowledge—but still these are forms of knowledge related to how things work rather than whether things concur:

> . . . a causal claim is not about a regularity between separate things or events but about what an object is like and what it can do and only derivatively what it will do in any particular situation (Sayer, 1992, p.105).

Sayer offers the example of a drunk who tries to discover the cause of his drunkenness by drinking whisky and soda one day, gin and soda the next, and so on, reaching the conclusion that the common factor (and hence the cause) is—the soda. We might opt for the alcohol rather than the soda, but

> . . . what gives such an inference credibility is not merely the knowledge that alcohol was a common factor but that it has a mechanism capable of inducing drunkenness (Sayer, 1992, p.115).

Whether that capacity to induce drunkenness is realized may depend on many circumstances—such as body weight, tolerance levels, and what else is being consumed as well as the amount of alcohol imbibed. Our understanding of conditions such as body weight is again based on how they work in mediating the effects of alcohol consumption. The recent controversy over recommended levels of "safe" drinking provides an apt example. The relation between alcohol consumption and mortality shows a U-shaped pattern, with higher mortality rates among those who consume none or a lot compared with those who are slight or "moderate" drinkers. The suggestion that alcohol when consumed in moderation can be health promoting remained controversial, however, so long as it was based only on evidence of an association between moderate consumption and lower mortality rates. The discovery of how various constituents in red wine can contribute directly to health (that is, the discovery of how it works) has been decisive in establishing (that is, making intelligible) a causal link between the two.

Sayer argues that the discovery of a regularity may or may not indicate a causal relationship, but it never establishes it. As we have seen in reviewing the methods of agreement and difference, regularity can prove a very poor guide indeed. By contrast, the discovery of an intelligible relationship between cause and effect provides a sounder basis for inference, which we can expect to last unless and until our understanding of how things work should prove inadequate (as usual) or (much less likely) mistaken. One of the most celebrated scientific observations of the century—the observation of light curving around the sun during the eclipse of 1919—illustrates the point very aptly, since the amount that the light was bent conformed exactly to Einstein's prediction of how things work (in this case, how light propagates) if space-time is curved. There was no need, and indeed no possibility, of repeating this observation: its impact lay not in the correlation of events but in the explanatory power of Einstein's relativity theory.

Our theories of how things work are not always so sophisticated (or complex) as Einstein's, but fortunately they need not be. In everyday life, a more prosaic working knowledge of causal processes is often enough to "get by" in our roles as producers, consumers, relatives, friends, or whatever. Take our role as voters in an election. One way or another an election produces change, whether governments or presidents survive or fall. We would not enter the poll booth if we do not have some working knowledge of how an election works (or perhaps even if we do). We at least understand the basics of the process if not the fine detail (such as a complicated system of electoral colleges) or why it produces any particular outcome. The exercise of causal power is rendered intelligible in this case through our understanding of the electoral process, regardless of any particular result. The latter depends on the specific conditions prevailing at the time. Or as Sayer puts it, "the relationship between causal powers or mechanisms and their effects is therefore not fixed, but contingent" (1992, p.107), as effects vary according to circumstance.

Making Causality Intelligible

- *In everyday life we rely on our practical knowledge of how things work*
- *Our own bodily experiences make causality intelligible*
- *Intelligibility is vital in distinguishing causality from correlation*
- *Intelligibility is established through establishing causal powers (and liabilities) rather than regular patterns*

This brings us back to the question of how we can distinguish causal factors from contextual conditions. One possible answer is that causal factors are necessary while contextual conditions are contingent. But necessary to what?

Sayer (1992, p.105) suggests that "particular ways-of-acting or mechanisms exist necessarily in virtue of their object's nature"—giving, as examples, planes (which can fly), gunpowder (which can explode), and multinational firms (which can buy labor cheap abroad). Planes can only fly, of course, if their engines work and their form is aerodynamic and, as these are improved, they can fly further and faster. Such powers or "properties" (to borrow an appropriate term from grounded theory) need not be static—they can change and evolve, just as "a child's cognitive powers increase as it grows" (Sayer, 1992, p.105).

Causal Powers

This view of causality tends to separate the identification of causes from the analysis of processes. Causality is invested in social structures, entities, actors, actions, behaviors and beliefs (this list is suggestive not exhaustive) rather than in the (contingent) connection between one event and another. Causal powers may initiate such processes (which may help in identifying them); but these powers exist independent of whether they are ever "activated"—just as my institution has the power to make me redundant, even if it never (I hope) exercises that power. One reason it may never do so rests in the existence of countervailing powers—for example, on the part of staff to oppose or resist redundancy. But even a causal power that is never exercised may still have a significant effect, arising from anticipation of the possibility (or perhaps probability) that it could be.

While causal powers are integral to entities and so on by virtue of their nature, the conditions that affect whether those powers are exercised, and with what consequences, are contingent—in that they may or may not be present. An institution, for example, is more likely to look for redundancies among its staff in a period of financial stringency or structural reorganization than during a period of expansion or incremental change. Its ability to push through redundancies may depend on the strength of staff opposition. This in turn may be strengthened or weakened by various conditions, such as levels of union organization, social security support, assistance with retraining and redeployment, or current rates of unemployment.

Not only the exercise of causal powers varies according to conditions, but also their very existence. To take a simple example, consider the way the properties of water change with temperature. Walking on water is not so difficult when the surrounding temperature is low enough to turn it into (thick) ice, and you can immerse (part of) your body in it when the surrounding temperature is high enough to turn the ice back into water—and, better still, warm it to something approaching 20 degrees! The low temperature is a necessary condition for water to have one property (as a solid); and a temperature high enough to melt the ice is a necessary condition for it to have the other (as a liquid). Incidentally, Sayer also talks of causal liabilities—which might refer, for example, to our own susceptibilities (given our thin skins) to changes in water temperature.

According to Sayer, a condition is either necessary, or it is not—there is no half-way house with necessity. Thus you cannot write (or read) a book without having a language—that is a necessary condition. You also have to find the time—another necessary condition. On the other hand, the conditions under which you write (or read) it may be a matter of choice—though here again, there may be some necessary conditions which dictate that choice: such as the invention of writing, the pen, the printing press, the typewriter or the wordprocessor. Many such conditions may be required for a simple causal power to be exercised effectively—think of my earlier example of what is involved in shopping in contemporary society. Also many causal powers may operate simultaneously—just as in a theatre, there are usually many things happening at once as the drama unfolds.

From this perspective, the analysis of causes and conditions requires first and foremost the identification of causal and conditional necessities. More simply put, it requires an analysis of how things work—which requires an understanding of both. In everyday life, a working knowledge may be sufficient, but a scientific approach certainly requires a more rigorous and penetrating approach to explanation. To switch on the light, you do not need to understand the behavior of electrons and ions—but you might have to if you want to design new materials for conducting electricity (see inset box).

How can we identify these causal powers and liabilities if not through regularities and the methods of agreement and difference? Sayer (1992, p.235) suggests that causal and conditional necessities can be explained "by reference to the structure and constitution of the objects which possess them." These in turn can be investigated by asking questions "which may seem simple to the point of banality" (Sayer, 1992, p.235) along the lines of "what is it about X that allows it to produce effect Y"? Such questions are easier to ask than to answer, though, since to answer them we must examine the internal relations and structure of objects that are typically overlooked in common sense thinking.

Take the family as an example. The term "the family" is, of course, an abstraction (and one with ideological overtones at that) since there is not one family but many families. Just as Lakoff argues in relation to "mother," the term "family (in everyday use) conveys a cluster of meanings that are "motivated" by idealized cognitive models. But this abstraction may serve a conceptual (as well as an ideological) purpose, in distinguishing the properties and relations that are constitutive of those social groups that we dignify with the description. Thus the term "family" can also serve as a shorthand to distinguish one entity (with its distinctive relations and structure) from others.

How Things Work in Scientific Explanation

Much scientific endeavour is devoted to deciding between rival explanations. Take a question which, according to *New Scientist* (May 7, 1997) is one of the "big" problems of contemporary physics: can high temperature superconductors conduct electricity without resistance because of pair-bonding of electrons or because of fluctuations in electron spin? If the answer is pair-bonding, then conductors should show "S" symmetry; if spin fluctuations, they should show "D" symmetry. A test was devised to detect (though indirectly) the type of symmetry shown. It showed "D" symmetry, and opinion swung behind this explanation of how superconductivity works. Meantime, those supporting the pair-bonding explanation have found further evidence that pair-bonding does occur between polarized electrons. Now opinion has swung back a little toward this explanation—though the smart money backs some combination of both explanations.

Another example in the same issue of *New Scientist* concerned the chance discovery of X-rays by scientists trying to design a portable proton accelerator the size of a microwave. They failed to produce protons, but to their surprise they did produce X-rays—at a much higher energy level than anyone thought possible. This puzzle—how does the would-be proton accelerator produce such high energy X-rays—has become the focus of their current research. Once understood, it may make possible the production of portable X-ray machines.

The details of these examples are unimportant—what counts is the insight they offer into scientific endeavor—the effort to understand how things work, whether high temperature superconductivity or would-be proton accelerators that produce high energy X-rays. In both cases, scientists advance theories to account for the available evidence and devise (if they can) tests that discriminate between rival theories.

Suppose we ask what it is about a group of people which merits this description? Do families have a particular set of relations of authority, such as we associate with parenting (young) children? Do they sustain a particular set of affective relationships, such as we associate with relationships between siblings or spouses? Do families have characteristic economic relations of production, distribution, and exchange? What about the "role" of the family in reproduction and socialization, or in the care of those afflicted by age or adversity?

However, in analyzing process we are less concerned with what families are than with what they do. What is it about families that makes them act in some ways and with some effects (but not in others?) The answers may be found in "how families work"—that is, in how they accomplish such activities as reproduction, socialization, and resource distribution. We all know that families can accomplish these things in different ways—reproduction within or outside of marriage, socialization in or outside the home, resources expropriated or shared, and so on. These differences in turn may reflect the conditional necessities associated with wider social structures, such as systems of education, taxation and social security, so that we cannot understand how families work in isolation from a network of wider social and economic structures in which families find themselves embedded. For example, technological advances in contraception may have significant effects on the possibilities of family formation and reproductive activity, while structural changes in employment may have a significant impact on how resources and responsibilities can be distributed within families.

We are dealing here, then, with a tangled web of relationships in which our primary concern is to identify how structural forces shape various (im-)possibilities for action. If married women are excluded from the labor market, for example, this may make certain kinds of resource distribution within families impossible, while facilitating others. Thinking about how things work can be illuminated by examining how certain actions are made possible (or facilitated) while others are made impossible (or constrained). But how can we identify and examine such (im-)possibilities? Though regularities may (or may not) offer (or obscure) some clues, ultimately we have to rely on reasoning (about how things work) informed by observation (of their effects). And the answers we are looking for are rarely transparent—we have to figure them out in light of the available data.

Take the following exchange (which I had with my son, who is working at the checkout of a supermarket) as an example:

> *Dad:* Quiet day today?
> *Son:* Yes, it was actually. It's near the end of the month, people haven't been paid.

Here we have an explanation that makes sense of why the supermarket was quiet—people have run out of cash and are waiting for payday. This explanation may have been suggested by observing a pattern here—if every month the same thing happens (and it only happens toward the end of the month). But the reasoning is concerned with structural conditions (payday is at the end of the month) and causal powers (if people do not have the money, they cannot shop).

These suggestions may take us some way beyond the initial boundaries of grounded theory, which has generally little to say about social structures and is mostly opposed to reasoning based on (im-)possibilities. This may be in part because in grounded theory process is conceived more with the dynamics of social interaction than with the implications of (social) structures. However, this is an issue that I explore more fully in the next chapter.

IN CONCLUSION

Any analysis of process requires some account of the nature of conditions and causes. In grounded theory the emphasis is very much on conditions, where we are offered the assistance of the conditional matrix and tracing conditional paths as frameworks for inquiry. Causal factors, on the other hand, figure in the analysis only in the guise of causal conditions.

Despite the central place of conditions in the analysis of process, the distinctions which Strauss and Corbin draw between different types of conditions are not very clear. The distinctions they do draw, for example, between intervening and contextual conditions, do not correspond to our usual use of these terms. I suggested that it might be helpful to consider contextual conditions as those within which an event occurs (rather than as properties of the event) and intervening conditions as those which do intervene in the temporal sequence between cause and effect.

I suggested that other distinctions between conditions also have an important role in the analysis of process. These include distinctions between necessary and sufficient conditions and those only contingently related to an event; between conditions which are single, multiple, or conjunctural; and between local and general conditions. One important reason for distinguishing between these different kinds of conditions is that they may require different types of explanation. In particular, the methods we use to establish single causation may be defeated by the problem of accounting for multiple and conjunctural causation. Glaser and Strauss recognize this problem and in defining the theoretical goals of grounded theory they explicitly disavow the aim of establishing causal generalizations, which they associate with analytic induction. The focus in grounded theory seems to be fixed on multiple and conjunctural forms of causation. However, the methodology of constant comparison to establish patterns of similarity and difference seems wedded still to the analysis of single causes. This seems so, even though the main most potent (though still limited) weapons of this methodology—the methods of agreement and difference—are not used in grounded theory.

In seeking an alternative basis for establishing causality, I suggested that in addition to temporal sequence, we can take account of the intelligibility of causal relations. This intelligibility is rooted in our bodily experience and practical (but also theoretical) knowledge of how things work. This approach to causality places less emphasis on correlation in establishing causal relations, though it need not discount regularity altogether, since this may still provide a basis (albeit limited and insecure) for causal inference. The focus, however, shifts to the causal powers and liabilities invested in the entities that shape our world—social as well as physical—and the capacity these have to produce particular effects. Whether they do produce such effects is a contingent matter; that they can do so is not. The entities that have these powers include, of course, ourselves, though not in isolation from other organisms or from the emergent powers of the social structures that we create through our interactions. This raises the general question of the role that structure and human agency play in the analysis of process. This is the main theme of the following chapter.

CHAPTER 8

Structure and Agency

The problem of relating structure and agency in accounting for process is by no means unique to grounded theory. The need to reconcile the influence of situational constraints with the freedom of human agency is a perennial problem of philosophy (in the form of debates about determinism and free will) as well as a pervasive problem in social theory (Giddens, 1984). In this chapter, we consider how structural factors are identified and used in analysis of process put forward in grounded theory.

To analyze the effects of structure on agency requires a consideration of time. In grounded theory temporal processes are conceived largely in terms of stages, though there is some disagreement over whether these stages—in the shape of a progressive movement from one stage to another—are an essential aspect of any analysis of process. The issue of stages also raises questions about the periodization of change, and the role of the analyst in constructing that periodization.

A more fundamental issue concerns the way the interaction of structure and agency is conceived in time. Are structural effects contemporaneous with and mediated by interaction or do they precede and constrain it? With its roots in symbolic interactionism, it is not surprising to find that grounded theory emphasizes interaction above all as the medium through which structural conditions are expressed and realized. An alternative though related approach suggests that structure both predates agency and is mediated by it. This approach takes more explicit account of the way in which structural effects can be both experienced as pre-established constraints and mediated by interaction through time. Here I lean heavily on the "morphogenetic" theory advanced by Archer.

Once the respective roles of structure and agency are more clearly distinguished, it is possible to identify the key features of their interplay. Here I again follow Archer's attempt to set out some of the main issues to be considered in any analysis of how this interplay proceeds. This account can be accommodated within the conditions-action-consequences framework of the coding paradigm, but it offers a much fuller picture of how such an analysis can be undertaken. This includes, for example, the identification of different situational logics or the differentiation between needs, roles, and interests in the analysis of interaction. I discuss Archer's analysis of process at some length, as it raises issues that can only help to sensitize us to the problems we confront in trying to analyze process. However, this does not imply we need to accept it uncritically.

STRUCTURE IN GROUNDED THEORY

To consider the role of structure (and agency) in the study of process, we can turn first to Glaser's account of how process is analyzed. I noted earlier that Glaser characterizes a process as having (at least) two "stages." Staging is not incidental to this conception of process, it is fundamental: no stages, no process. For example, Glaser cites the mistaken identification of "shifting" as a basic social process:

> . . . in one study, "shifting" was seen as a BSP [basic social process]. After review we found no stages and reconceptualized it as a basic social structural condition—shifts—confronting people and organizations that had a twenty four hour a day operation (Glaser, 1978, p.97).

Thus basic structural conditions are not conceived by Glaser as part of the basic social process but rather as (pre-)conditions that agents "confront."

This is not because social structures are considered static, as they too (we are told) can be analyzed in terms of process—that is, in terms of a transition from one stage to another. Glaser distinguishes between psychological and structural processes through examples:

Basic social psychological processes (BSPP)	*Basic social structural processes (BSSP)*
Becoming, highlighting, pesonalizing, health optimizing, awe inspiring, etc.	(De-)bureaucratization, routinization, (de-)centralization, organizational growth, admitting or recruiting procedures, succession, etc.

Structural processes also involve stages, just as promotion from one post to another can represent stages in an organizational career or transition from one class to another can represent stages in an education process. This conception of structure as process is contrasted by Glaser with more static accounts, which simply refer to a (changing) set of structural conditions without analyzing the process of structural change. However, Glaser seems unsure whether such an analysis is really required rather than simply "focusing on one [process] and using variables from the other" (Glaser, 1978, p.102):

> Perhaps the BSPP [psychological process] is more prevalent and relevant to understanding behavior, since one does not need the BSSP [structural process] to understand it, but usually one needs a BSPP [psychological process] to understand the focus on a BSSP [structural process (Glaser, 1978, p.102).

Although "the most sophisticated sociological renditions" involve both, this "takes skill and clarity of purpose" and Glaser suggests that analysts may opt to emphasize one at the expense of the other (though they should be explicit in doing so). The balance of emphasis may depend on whether structural change is rapid or slow, though again psychological processes seem to be paramount. Thus Glaser argues that psychological processes are likely to be emphasized where change is rapid but also where there is little change:

> In studying a process which optimizes change . . . it is likely the emergent mix would emphasize the BSPP [psychological process]. . . In studying a phenomenon that requires little change . . . structural process might not be as important . . . (Glaser, 1978, p.103).

Even where the focus of study is on a "structural phenomenon as it is growing" it is suggested only that "one would also bring in the new BSSP [structural process] that supports the BSPP [psychological process]" (Glaser, 1978, p.103). Although we are told that the issue has to be settled "empirically for any particular study" (Glaser, 1978, p.102) there is little doubt where Glaser's inclination lies—that basic social process "implies a BSPP (that is, a social psychological process) unless otherwise stated. This is certainly consistent with the presentation of grounded theory as first and foremost concerned with action and interaction.

A similar emphasis is apparent in the account presented by Strauss and Corbin. Here process is conceived as a sequence of interactions toward a desired goal—a characterization consistent with the persistent emphasis on purposeful or strategic action in grounded theory. Process is described as "a way of giving life to data by taking snapshots of action/interaction and linking them to form a sequence or series" (Strauss & Corbin, 1990, p.144).

STAGES

Thus structural conditions are recognized in grounded theory, but their role tends to remain incidental rather than integral to the analysis of process. This is so even where a structural process is itself the focus of study. Moreover, structural conditions only figure in the analysis insofar as they can be considered as processes demarcated in terms of stages. As this demarcation can be accomplished by the analyst, for "purely heuristic" reasons, this restriction is not as severe as it might be if such stages have to be in vivo, that is, recognized as such by the persons involved (Glaser, 1978, p.98). However, the restriction is severe nonetheless, since it rules out analysis of any structural conditions or processes that do not conform to this prescription.

Moreover, as stages are conceived of largely in terms of whether organizations are expanding or contracting, this represents a further restriction on what can be included in the analysis of process. We can include a social structure such as a bureaucracy, for example—but only if it is growing (or diminishing). However, social structures change in more ways than through expansion and contraction. They may change in terms of their internal structure and relations, without changing in size at all. For example, an organization may change its procedures for recruitment and unfair dismissal, perhaps in response to new legislation, without altering the size of its staff. An educational institution can change its methods of teaching and assessment or a health agency its methods of organizing a waiting list or of providing care. Such changes may have implications for size, but need not. Also, these changes may be rather resistant to conceptualization in terms of "stages."

According to Glaser, a stage must not only have a time dimension—that is, a perceptible beginning and end; but it must also "differentiate and account for variations in the problematic pattern of behavior" (Glaser, 1978, p.97). In other words, stages are not to be identified or defined in their own right, but rather with regard to their effects on the main phenomenon being studied. Otherwise they are not deemed relevant to the analysis.

This may seem a rather arbitrary criterion for distinguishing stages if we allow that a process of structural change may exhibit stages regardless of whether these stages have other effects. Some processes may not be amenable to conceptualization in terms of stages, for example, if they exhibit more continuous or incremental forms of change. Significant change does not only or always occur through phase transitions—as we can readily recognize by virtue of the remorseless process of aging. Groups and organizations also "age" (not just individuals) and this too may involve significant changes that are not necessarily conceptualized into recognisable stages. They may gain in vital experience and confidence or they may suffer an erosion of energy and commitment over time without any tangible threshold to mark a transition from one phase from another.

I am not denying for one moment the heuristic value of analysis in terms of stages—or what we might call "periodic" analysis. Moreover, I am happy to accept that periodic analysis may be more than a heuristic tool, since, as Glaser suggests, the stages, "phases," or "periods" that we (or our respondents) recognize may have an empirical reality. Thus the millennium, for example, is not a mere fantasy or fabrication but has a tangible reality in terms of social action—most notably in the projects undertaken to celebrate or exploit the event. Similarly, distinctions between prewar and postwar periods are not arbitrarily imposed on events if they reflect real changes in structures, attitudes and behavior in light of the experience that war generates. However, the question whether periodic analysis is an appropriate or useful way of analyzing process cannot be determined a priori. It may be that for some processes, periodization obscures more than it reveals.

Glaser's Analysis of Process in Terms of Stages

- *Recognizes structural "constraints" on action*
- *Emphasizes structural change as a process*
- *Focuses analysis on relevant stages*
- *Can consider stages as either heuristic or "in vivo"*
- *Tends to limit stages to particular processes (e.g., expansion)*
- *Discounts gradual or continuous changes that are not periodic in form*

Once we have freed structural conditions from such needless restrictions regarding stages, we can consider afresh how to incorporate structure into the analysis of process. The coding paradigm, the conditional matrix and the tracing of conditional paths are all intended as heuristic guides to aid this analysis. As we have seen, they suggest a sequential ordering of analysis (conditions—interaction—consequences), a range of factors to be considered (in the conditional matrix), and the need for focus (in tracing conditional paths).

Sequential ordering in grounded theory seems to require a historical and biographical approach to analyzing process: interactions can only be understood in terms of the specific (structural) conditions in which they occur and the particular (structural) consequences to which they give rise. Temporality is therefore an integral aspect of analysis, whether it is the structural conditions themselves that change over time or the interactions that they condition. Or, rather, process in grounded theory can be analyzed in terms of an evolution from conditions through interaction to consequences, which can in turn become new conditions.

On the other hand, the conditional matrix, as presented by Strauss and Corbin in a series of concentric circles, is clearly atemporal in conception.

There is no ordering in terms of prior conditions and subsequent consequences; if anything, analysis seems to proceed on the basis of simultaneity. This concern with current conditions is reinforced by the discussion of tracing conditional paths which identifies wider conditions of interaction (as in the example regarding the missing gloves) with no more than incidental regard to temporal considerations. This point is acknowledged explicitly by Strauss and Corbin:

> Temporality is built into the conditions. However, when we analytically stop the action/interaction to examine them, we view them cross-sectionally. We see them artificially, so to speak, as a slice of time, rather than over time, with relevant pasts, presents and futures (Strauss & Corbin, 1990, p.161).

In other words, the focus is on "existing" rather than "emergent" conditions, and the time dimension is incorporated into the present. However, we have already seen that temporal sequence is an essential aspect of the analysis of causality and conditions.

TEMPORAL SEQUENCE

If we were to incorporate a temporal dimension in the conditional matrix, we could separate out conditions, interactions and consequences as in Fig. 8.1.

In Fig. 8.1, conditions are prior to (interaction) but to some extent coincidental with it. There is an overlap rather than a sharp break between the two, whereas a more conventional account might present conditions are entirely antecedent (Fig.8.2).

The conventional account implies that structural conditions are antecedent, and therefore at the point of interaction they are entirely absent from the evolution of process. This view is represented in grounded theory in

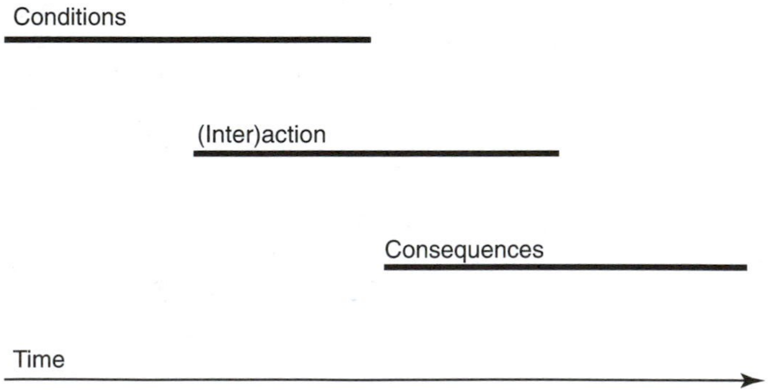

FIGURE 8.1 Temporal sequence in analyzing process (adapted from Archer, 1995, p.194).

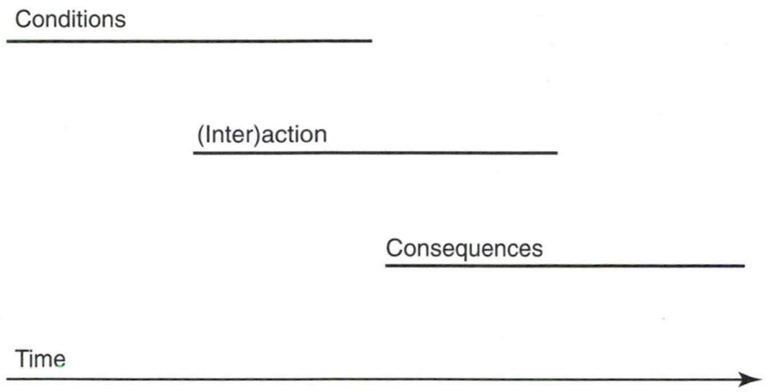

Conditions

(Inter)action

Consequences

Time

FIGURE 8.2 Conventional view of temporal sequence exhibited by process.

the assertion that causal conditions are antecedent. But Archer (1995, p.158) suggests there is no point in time (such as "the present") when only interaction matters and conditions are somehow suspended. Instead, there is an interplay between actors or agents and the social structures they inherit. This interplay is also represented in grounded theory in the recognition that conditions only affect outcomes through the mediation of interaction. We can map this as in Fig. 8.3, which shows conditions operating through interaction. That is, conditions are concurrent with (inter)action rather than preceding it.

However, this approach takes no account of the pre-existing character of initial conditions.

The way Archer presents process incorporates and integrates both these perspectives. Conditions precede interaction and therefore we have to accept them as given—we cannot alter them to suit ourselves. But while we cannot alter the conditions we inherit (at the time we inherit them), we can alter their effects (and so the influence of the conditions themselves) through in-

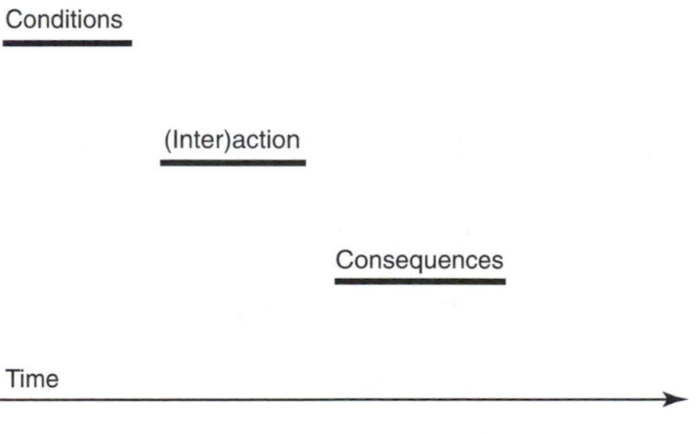

FIGURE 8.3 Analysis of process as a sequence of interactions.

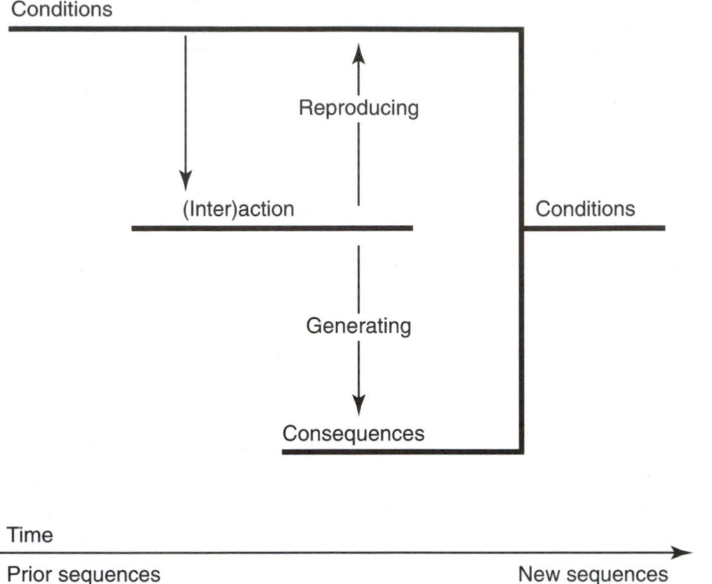

FIGURE 8.4 Reproduction and transformation in process (adapted from Archer, 1995, p.158).

teraction over time. The influence that conditions exert on outcomes (and the next cycle in the process) is therefore mediated by interaction.

One aspect of this interplay involves the reproduction of pre-existing structures. Another involves the generation of consequences, which become the preconditions of a further sequence (Fig. 8.4).

This approach incorporates a conventional insight—that social structures are dependent for their reproduction on interaction while that interaction also generates social change that can, in turn, facilitate or circumscribe further action—or, as Marx put it, people make history, but not in circumstances of their own choosing.

STRUCTURE AND AGENCY

Archer calls her approach "morphogenetic" (1995, p.75), to express the reshaping of forms through the interplay between structure and interaction. This expression is not only "unlovely" (which she acknowledges) but also inappropriate (which she does not). In any analysis of process, we have to consider morphostasis (reproduction) as well as morphogensis (transformation).

The interplay between conditions and interaction thus depends on a conceptual separation of structure and agency, since this is essential if we are to allow that conditions "have autonomy from, are pre-existent to, and are causally efficacious vis-à-vis agents" (Archer, 1995, p.60). If structure and agency are seen as "mutually constitutive," so that conditions are seen as entirely dependent on current interaction—and therefore always open to transformation—then they cannot be conceived as having an autonomous impact. Activity both predates and postdates the emergence of structures so that we can consider both the "analytic histories" of emergent structures and the way people either reproduce or transform those structures. Because they preexist interaction, emergent structures are not created by current activity, but only reproduced or transformed by it.

Archer also rejects the idea that agency and structure are mutually constitutive because this overburdens agents with responsibilities they cannot bear, such as excessive "knowledgeability":

> . . . people do not have and cannot attain "discursive penetration" of many unacknowledged conditions of action . . . agents have differential knowledgeability according to social position; and some agents have defective, deficient and distorted knowledge owing to the cultural manipulation of others (Archer, 1995, p.252).

Of course, defective knowledge can derive from many other factors than cultural manipulation; we often cannot explain why we act as we do. People act out of motives and interests that they may or may not be able to articulate.

How does this account compare with the conception of process in grounded theory? Archer suggests that "symbolic interactionism" generally fails to separate structure and agency:

> . . . in viewing entities such as social institutions as purely dramatic conventions which depended upon co-operative acts of agents in sustaining a particular definition of the situation, Symbolic Interactionists in particular elided "structure" and "agency" in three key ways . . . (i) a denial of their separability, because, (ii) every aspect of "structure" is held to be activity-dependent in the present tense and equally open to transformation, and (iii) the conviction that any causal efficacy of structure is dependent upon its evocation by agency (Archer, 1995, p.60).

As grounded theory largely has its roots in this tradition, its problems of dealing with process may ultimately be traced to this elision of structure and agency. This elision is evident in the way Strauss and Corbin present process (Fig. 8.5), in which changing conditions impact on interaction only insofar as they are themselves reproduced through action.

In contrast to Archer's account, conditions do not figure as autonomous, prior and causal, but rather as dependent on interaction, simultaneous with it, and not causally efficacious in their own right.

If structure and agency are not separated, Archer argues, it becomes impossible to examine the interplay between them since:

> causation is always the joint and equal responsibility of structure and agency and nothing is ever more attributable to one rather than the other, at any given point in time (Archer, 1995, p.64).

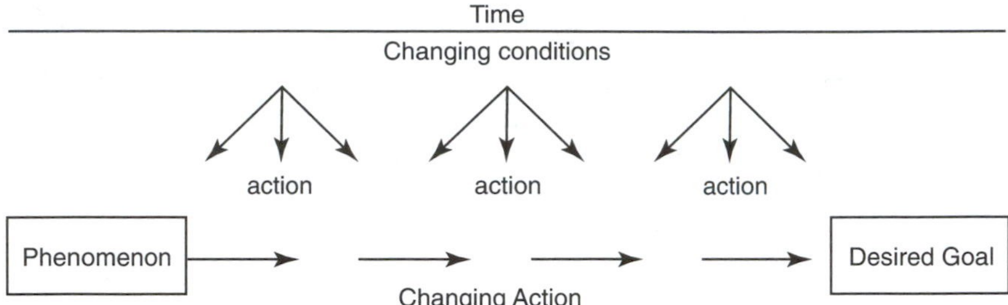

FIGURE 8.5 Strauss and Corbin's picture of process (adapted from Strauss & Corbin, 1990, p.145).

This joint responsibility is evident in the conditional matrix, where no differentiation is made between structural conditions (distinguished at various levels) and interaction.

The conditional matrix does distinguish conditions at different levels, from local to international. These distinctions are initially drawn in terms of distance from an interactional sequence, with the outer circle deemed "most distant." Distance here presumably refers to distance between a condition and the interaction that it affects. However, the authors also suggest that research can focus on any level—so long as all the others are taken into account:

> . . . the researcher needs to fill in the specific conditional features for each level that pertain to the chosen area of investigation. . . Regardless of the level within which a phenomenon is located, that phenomenon will stand in conditional relationship to levels above and below it, as well as within the level itself (Strauss & Corbin, 1990, p.162).

While this arrangement of conditions makes some sense in terms of levels of generality, it may be misleading to equate differences in generality with differences in "distance" between condition and effect. We can readily recognize conditions (most obviously war and peace) where international action may have an immediate and direct effect on interaction(s) without needing any meaningful mediation through intermediate conditions. We may also doubt whether conditions at one level are "bounded" by higher levels or themselves set boundaries for lower levels.

Take, for example, the relationship between identity and interaction. I am variously a son, brother, husband, father, and friend; an academic; living in Edinburgh; a Scot; a citizen of the UK; a European; and so on, etc. These various "identities" certainly condition my interactions, but my being a Scot (for example) does not stand in any less direct relation to my interactions than my being an academic. And of course, being an academic is itself being part of an international community, so levels overlap; while being a Scot can conflict with being a member of the United Kingdom, for conditions at different levels may also conflict.

As a guide to analysis, the conditional matrix therefore has some limitations, though it does at least address the problem of how we can connect conditions to interaction. There may, however, be some other useful distinctions than those concerned with the generality of different conditions. Since I have followed Archer in separating structure and agency, let us look now at how she considers the interplay between conditions and interactions over time. As her approach conceptualizes issues that remain at best implicit in grounded theory, it is worth discussing her account in some detail.

GENERATIVE MECHANISMS

Archer summarizes the main tenets of this evolutionary perspective as follows (1995, p.159):

- Pre-existent structures act as generative mechanisms
- These mechanisms interact with other objects possessing causal powers and liabilities proper to them in what is a stratified social world
- Nonpredictable but explicable outcomes arise from these interactions

What does it mean to claim that structures act as generative mechanisms? To illustrate, Archer (1995, p.174) cites a demographic structure, that is no mere aggregate of people, she argues, but a structure that can and does modify the powers of people to change it. For example, a demographic structure defines the size of the relevant group of child bearing couples whose reproductive behavior could transform that structure. Likewise, the emergence of a "top-heavy" demographic structure (associated with an aging population) has implications for the levels of taxation required to fund pensions (other things being equal). Thus such demographic structures have "generative powers" — in these examples on levels of reproduction and taxation—even if their effects may be obscured by other factors (such as pensions less generous than they otherwise might be).

Structural properties are emergent when structures develop capacities to affect interaction in particular ways. Simply put, you cannot shop without shops—at least unless and until new structures emerge, that allow virtual shopping. Such emergent properties are not to be confused with "an overt and relatively enduring patterning in social life" (Archer, 1995, p.172) achieved simply by aggregating and classifying institutions, organizations, or groups on the basis of common features or regularities. In Archer's view, these taxonomic categories (such as "the health service") refer to observable phenomena that are "contingent combinations of disparate elements from different strata which happen to co-manifest themselves at a given time" (1995, p.173). An emergent property is not just a juxtaposition of heterogeneous features (such as those with some connection to health) but exhibits a homogeneity based on "the natural necessity of its internal relations" as "what the entity is and its very existence depends upon them" (Archer, 1995, p.173). Some relationships of an entity are external and contingent, but those which are constitutive of its emergent properties must be internal and necessary. We might say, in a term which captures both these qualities, that some properties are "inherent."

On this basis, the relationship between dentists and doctors is external and contingent, though both may be classified as part of the health service. But a relationship between doctor and patient is not external and contingent but inherent (internal and necessary)—because the one cannot exist without the other. There can be no doctors if there are no patients, and there can be no patients if there are no doctors. The emergence of a medical relationship between doctor and patient is a "generative mechanism" with implications for how health (and especially illness) is conceived and managed. This "internal and necessary" relationship itself requires others, such as medical training—as well as the rise of professional authority (in place of patronage), the development and diffusion of scientific and technical expertise, the acquisition and distribution of resources capable of sustaining medical interventions, and the general medicalization of health issues. Whether such properties emerge is contingent—some societies, for example, have developed ways of distributing health resources that allow (largely) free access to health care at the point of delivery, while others have not. And the impact of emergent properties may depend on other factors that are contingent and external—just as poverty and discrimination can affect access to health services.

Moreover, outcomes are the product of an interplay between emergent properties (such as free access to health care) and their reception by people "with their own emergent powers of self and social reflection." Outcomes are identified:

> by analyzing the generative mechanisms potentially generated from structures (and cultures) as emergent properties and their reception by people, with their own emergent powers of self and social reflection. Outcomes never simply mirror one or the other, but are products of their interplay (Archer, 1995, p.175).

Thus the doctor–patient relationship is an emergent property (albeit one with a long history) but its generative power may depend on whether and how far people are prepared to fulfill (or resist) the doctor and patient roles. This may be undermined variously by ignorance and incapacity, by degrees of apathy or commitment, or by education and access to alternative sources of specialist expertise (such as the Internet). In considering how the interplay between conditions and interaction affects outcomes, we have to consider not just individual actors but the emergent powers of different groups and agencies (such as professional and patient organizations).

Archer distinguishes between different types of emergent property:

> In society there are a variety of emergent properties—structural, cultural and agential, each of which is irreducible to the others, has relative autonomy, and is also relatively enduring (Archer, 1995, p.175).

Whereas the conditional matrix distinguishes conditions only according to distance, Archer distinguishes emergent properties in terms of material, logical, and agential powers. Let us consider each in turn.

Structural Properties

Structural properties are distinguished by their "primary dependence upon material resources, both physical and human" (Archer, 1995, p.175). Structures may be legitimated by ideas, but their causal efficacy does not depend on them, for it can ultimately be sustained by force and manipulation. Structural properties ca have effects independent of the interpretations placed on them, as when failures in food production (or distribution) precipitate famine or the imposition of fees exclude those who lack the requisite resources. Archer here is emphasizing the material aspects of causal efficacy. We can see that schools require classrooms, universities require libraries, and hospitals require beds. We may think of admission to hospital or university as a social act, but it is also a physical one.

In grounded theory this material context may be easily overlooked through the overriding focus on intention, action, and interpretation. Archer's emphasis on physical structures and their implications for enforcement and constraint may provide a useful corrective to an account of process preoccupied with how people interact.

Cultural Properties

A cultural property refers "to all intelligibilia," by which Archer means "any item which has the dispositional capacity of being understood by someone" (1995, p.180). She offers the example of a soufflé recipe. The logical status of a cultural property is not dependent on our ability to conceive, comprehend or use it, she argues, but rather on the consistency of its propositions—both among themselves and with other ideas:

> Analytically, at any given point in time, the items populating the [cultural system] realm have escaped their creators and have logical relationships among one another which are totally independent of what people know, feel or believe about them (Archer, 1995, p.182).

However, as Archer (1995, p.180) acknowledges, "we do not live by propositions alone" but also by myths and mysteries, tastes and prejudices, empathies and animosities—which tend to cloud our judgment and derail our logic. But all this "takes place beyond or outside the canons of logic" and is therefore conceived as an aspect of "socio-cultural" interaction. In such interaction, causal rather than logical relations may rule—just as a committee is liable to follow its most forceful member rather than its most logical path. Whereas the connections between ideas are defined by necessity, those relating ideas to interaction are contingent.

It is not clear (at least to me) why Archer insists that emergent cultural properties must be defined by relations of logical consistency. Only mathematics would seem a likely candidate to meet this requirement—and Gödel's theorems established that even in mathematics (and logic) any system is necessarily incomplete and cannot be proven to be logically self-consistent (Barrow, 1993, p.119). In science, ideas are not taken up because they are consistent but because they can explain (well enough) how things work. Thus two of the major theories of this century, the theories of relativity and quantum mechanics, are inconsistent—but each works well enough in its own sphere, of the immense or the minuscule. Moreover, an appeal to the "universality of the law of non-contradiction" (Archer, 1995, p.180) takes no account of the development of multivalent logics—most notably "fuzzy logic" with its direct challenge to the laws of Aristotelian logic. In reducing meaningful statements to propositional logic, Archer also ignores the metaphorical, image-schematic and metonymic forms of meaning explored by Lakoff.

The prevalence and persistence of various forms of religion, based on faith rather than logic, would seem to require an account of "cultural properties" that is more encompassing than the rather cold world of propositions that Archer offers. The "force" and "durability" of an idea need not lie in its consistency, as Archer herself (ironically) demonstrates in bemoaning the persistence of methodological and ontological inconsistencies in social science. Faith is sufficient for causal efficacy, no matter how misconceived the object of our beliefs may be—as the suicides of members of a religious sect intent on liberating the aliens possessing their bodies sadly demonstrate.

The implied separation of cultural properties from material resources is also open to criticism. The emergence of pop music, for example, would seem to have a material foundation in the construction and widespread distribution of transistor radios, record players and later forms of hi fi—not to mention the economic and social impact of growing affluence and longer schooling on youth culture. The history of any branch of music (such as the blues) may identify material and social factors that have had a tangible impact on its evolution. The rhythmic tempos of some folk music may be related quite closely to the pace of the production processes that it accompanied. The cultural predominance of Hollywood may be a product of the financial exigencies of the film industry. The aesthetic (and not just the economic) value of a "priceless" work of art can hardly be separated from its sentimental associations, its scarcity value, its attraction as an investment, and its role as a symbol of wealth (or at least, conspicuous consumption). This is a sad truth for artists, since it means that only after death may the full value of their work be realized.

Nevertheless, Archer's distinction between the logical and causal force of "cultural properties" does emphasize dimensions of analysis that might otherwise be overlooked in grounded theory. The cultural context can be analyzed as a given, at least in the sense that the evolution of ideas (in the broadest sense of cultural artifacts) has a profound impact on the logical possibilities and interpretive perceptions of actors.

Agency

The third "emergent power" that Archer recognizes is that of agency. Here she distinguishes between individual persons, social agents and social actors, regarding "all three as indispensable in social theorizing but as irreducible to one another" (Archer, 1995, p.249).

By "persons" Archer refers to the defining properties of people that are necessary conditions for any kind of social life. Foremost among these is a continuous sense of self over time, as without it social activities that take place over time—like hoping, preparing, planning, and so on—would be inconceivable. This sense of self is universal—a sine qua non of being human—and not to be confused with cultural concepts of self that have evolved historically, embodied, for example, in legalistic conceptions of "what is a person." The sense of self, Archer argues, is indispensable to our experience as thinking, embodied beings who can make distinctions—including those between the physical and the social. Because the self is embodied, our embodiment as human animals both defines who can be persons, what capacities we can develop and what we can do. This is why some social conditions can be judged "dehumanizing"—if they do not meet the basic human needs required in order to realize our natural capacities. In short, we always have to consider our next meal. *What* we eat as a meal may be socially mediated but that we eat is not. As animals, we have first of all to survive—and reproduce—in order to be socially active. Hence the "cumulative experiences of our environment will foster propensities, capacities and aversions" that shape the social identities we later acquire.

As well as "persons" we can also be "actors": we can acquire a social as well as a person identity. Such as social identity is acquired "by investing ourselves in a role and personifying it in a particularistic way" (Archer, 1995, p.256). Not everyone, Archer argues, acquires such a social identity:

> Everyone has a personal identity, but each does not, I submit, have a social identity, that is any role in which they can invest enough of themselves to feel at home with what they have become (Archer, 1995, p.256).

There are social roles that are occupied by default (such as "the unemployed" or "the homeless") and that deny rather than express our human capacities.

Finally, we act as "agents," but agents of what? People are agents "of the socio-cultural system into which they are born" and of "the systemic features they transform" (Archer, 1995, p.257):

> Everyone is born into an ongoing socio-cultural system and all have agential effects on stability or change, if only by merely being within it—physically and numerically. Moreover, the world, structured as they find it and are placed in it, is the one in which they live and move to have their social being: yet there is no being without doing and no doing without consequences (Archer, 1995, p.259).

Thus we are agents because to live we must act, and our actions have (social) consequences. But Archer adds that we act as agents only as members of "collectivities"—of groups that share interests and life chances. Agency is therefore concerned with the articulation of shared interests, the organization of collective action, the generation of social movements and the exercise of influence over decision-making (Archer, 1995, pp.259–260). Here Archer distinguishes between "corporate" agents—who recognize and actively promote their vested interests—and "primary" agents—who do neither, but merely react and respond through unarticulated and uncoordinated forms of interaction. Moreover, some agents may influence events simply by virtue of their existence, while others may face constraints that curtail their capacity to act otherwise. Thus (and paradoxically) we can be "agents" without acting, since sometimes we cannot, and sometimes we need not, act—but in either case we may still "make a difference" to stability or change (Archer, 1995, pp.118–119).

Different Kinds of Agency in Interaction

- *Persons—whose needs must be as a condition for any kind of social life*
- *Actors—who invest in social roles and identities*
- *Agents—who collectively share interests and life chances*

The details of these distinctions—between persons, actors, and agents—can certainly be challenged. Archer clearly has no time for a "great man" view of history, where individuals can act as agents in their own right and not simply as part of collectivities. History in her book is made by classes, not by kings. Yet power is sometimes concentrated in the hands of individuals whose projects make a difference yet may or may not correspond to those of some collective interest. To exclude altogether the capacity of individuals qua individuals to act as agents of stability or change seems needlessly restrictive. It seems equally restrictive to confine the term actor only to those with "positive" social roles where people "may feel at home." We can act out social roles (such as our occupational roles) regardless of whether we feel (un-)comfortable with the social identities they may confer.

Despite some difficulties, these distinctions nevertheless offer a more differentiated view of agency and interaction than the approach in grounded theory, which tends to focus on our social roles as actors. In Archer's account, agency is not uni-dimensional but rather stratified into three dimensions: we act as individual selves (with needs), as social actors (with roles), and as social agents (with interests). This directs our attention to aspects of interaction we might otherwise overlook (or ignore)—notably the impact of our environment (on needs) and the effects of resource distribution (including power) on the articulation and pursuit of interests. From this perspective, a grounded theory is one that considers whether and how individuals (themselves or as part of collectivities) seek to satisfy their needs and pursue their interests as well as how they realize their social roles.

CONSEQUENCES AND STRUCTURAL ELABORATION

Consequences are defined by Strauss and Corbin as the "outcomes or results of action and interaction" (1990, p.97); which, they add, may "not always be predictable or what was intended" (1990, p.106). Also they note that "consequences of action/interaction at one point in time may become part of the conditions in another" (Strauss & Corbin, 1990, p.106). In suggesting that outcomes can be attributed entirely to interaction, Strauss and Corbin appear to discount the effects of structural conditions on process. On the other hand, they (implicitly) reinstate these conditions in suggesting that process results in consequences that then condition further interaction. How can we reduce this tension between the autonomous role of conditions and the mediating effect of interaction in grounded theory?

In Archer's view, although the causal effects of pre-existing structures are mediated by agents, this does not imply an arbitrary freedom to choose any course of action. We find ourselves born into or "placed" in social positions that circumscribe our freedom of action—resources, roles, and rules constrain or enable (interaction), and cannot be ignored. We do not have to conform, but we may pay a high price for nonconformity: altering situations "is not a matter of untrammelled choice but of confrontation and extrication which carry costs" (Archer, 1995, p.202). For example, we may choose not to marry, to avoid the legal, financial, moral, and juridical obligations it entails, preferring instead either to remain single or to cohabit. But cohabiting too is a choice involving situational constraints (related to child custody and maintenance, for example), while "opting for single or celibate status is not to opt out of situational constraints, but to be embroiled in another set" (Archer, 1995, p.202).

Vested Interests and Opportunity Costs

The major effect of "involuntaristic placement," Archer suggests, is "to endow different sections of society with different vested interests" which are not episodic but "both systematic and enduring" (Archer, 1995, p.203). Such vested interests are objective features of positions predisposing their occupants to "different courses of action and even different life courses" (Archer, 1995, p.203). Thus "vested interests are the means by which structural (and cultural) properties exert a conditional influence on subsequent action" though that influence "depends for its efficacy upon them being found good by large numbers of those share them" (Archer, 1995, pp.203–5). This is not simply a matter of interpretation, for real opportunity costs are associated with refusal to pursue vested interests that confer relative advantage; or as Archer puts it, "virtue carries a price tag."

Opportunity costs differ depending on one's position in a distributional structure—attaching different costs for the same interaction (such as marriage) and thereby also affecting which projects are entertained by people in different positions. For example, those in disadvantaged positions may be deterred from marriage altogether by the higher costs they may incur. They are not "forced" by conditions to reject marriage, but the premiums and prices associated with marriage may create good reasons for doing so. This in effect curtails the degree of interpretive volatility through which agents can assess preferential courses of action. Therefore Archer suggests that different degrees of interpretive freedom may be associated with particular positions, though constraints never fully determine the course of interaction. We can give up paid employment, allow our homes to be repossessed, and do voluntary work for a pittance—though we must (and, indeed, may) be willing to pay the price for our virtue.

There are some additional points worth noting.

First, vested interests may be promoted by various courses of action, requiring judgment of potential risks and opportunities. Thus the impact of conditions on action is not habitual or routine. Agents may attempt (un-)successfully to overcome situational constraints. And to exploit situational enablements, they may have to act in innovative ways, whether to catch up, stay put, or forge ahead.

Second, agents are not conceived as economic calculators concerned solely with material benefits and costs. Agents take account of other values, notably their moral repute. Archer (1995, p.211) suggests that they tend to balance their material and "ideal" interests, a balance in which agents generally try to avoid paying heavy material costs for their ideals, but take these ideals seriously into account in judging how to promote their material interests. Such ideals, of course, may themselves be shaped by situational constraints and enablements, reflecting the cultural concerns of religious sects, professional organizations and other "moral" communities.

Third, the differential costs associated with different positions may create vested interests in either reproducing or transforming the status quo. Although we tend to associate vested interests with those who benefit from their positions, Archer means any interests that are vested in a position, whether or not the position confers resources and power. Those who benefit from an existing structure have a vested interest in defending it, while those who fail to benefit (or lose out) have a vested interest in opposing it. Those benefiting therefore have an interest in reproducing the status quo, and those losing out have an interest in transforming it.

The Role of Vested Interests in Interaction

- *Vested interests may be promoted more or less effectively, or not pursued at all*
- *Agents take account of moral values as well as material benefits*
- *Vested interests dispose some to defend and some to challenge the status quo*

Situational Logics

In Archer's account, reproduction and transformation are processes mediated by the projects taken up by those occupying different positions. Such projects may be taken up or rejected depending on the premiums and penalties associated with them, and these in turn may depend on what opportunities for action (if any) are available. The basic idea here is to identify the logic of situations and its implication for action. Some circumstances call for opportunistic alliances, while other circumstances may call for defensive action and protecting one's corner. Archer distinguishes four different kinds of strategic action—"protection," "opportunism," "compromise," and (in decidedly uncompromising language) "elimination." She tries to explain the logic of opting for one strategy over another in terms of how closely different interests are articulated—whether the relationships between the various parties are necessary or contingent and whether their interests are compatible or incompatible (see Table 8.1).

TABLE 8.1 Situational Logics Associated with Relations between Emergent Properties

	Necessary	Contingent
Compatible	Protection	Opportunism
Incompatible	Compromise	Elimination

Source: adapted from Archer (1995, p.218).

For example, the caste institutions of ancient India, where economic, political, religious, legal and educational institutions were highly interconnected, created general "negative feedback loops which discouraged alterations" (Archer, 1995, p.220).

A now classic (and rather more accessible) examination of interplay of structural interests and situational logics is in terms of a game theory situation called "the prisoner's dilemma." This involves two prisoners, held separately, who have to choose whether to inform or remain silent. The situation can be structured to create an incentive to inform. If one informs, that prisoner goes free and the other suffers the maximum penalty. But if both inform, both receive a reduced sentence. Of course, if neither inform, then both go free. The dilemma concerns whether to cooperate with the other prisoner (and keep silent) or to defect. Computer simulations of this dilemma that allowed learning through repetition found the strategy that works best in the long run (so far) is also the simplest: "tit for tat" (developed by Anatol Rapoport of the University of Toronto). In this strategy a prisoner starts by cooperating and then responds in kind to whatever the other prisoner does. One reason this strategy works may lie in its very simplicity—it is possible for the other prisoner to figure out what is going on (Waldrop, 1994, pp.262–3). The virtue of this approach is that it recognizes that strategic interests may not always be clear-cut, but instead dependent on the strategies pursued by other actors (facing similar dilemmas) and on how these strategies evolve over time.

I do not intend to assess in any detail the merits of Archer's examples or, indeed, to assess in any detail her association of particular situational logics with different structural configurations. However, we can see in this approach, with its identification of different strategies and their explanation in terms of the logic of underlying situtions, as a powerful analytic tool for exploring the connections between structure, agency, and consequences.

END-POINTS

In Archer's account, process occurs in cycles that culminate in structural elaboration—the reproduction or transformation of structure through the interplay of structure and agency. She identifies three sources of structural elaboration in “the confluence of desires, power-induced compliance, and reciprocal exchange” (1995, p.296). Basically, this focuses on the social distribution of resources among different interests and the relations of power and exchange between them. The ability of some groups to shape interaction in accord with their own vested interests may vary with the degree to which vested interests are superimposed, concentrating bargaining power in relatively few hands. However, transactions are influenced not only by the initial bargaining positions of different interest groups, but also by their actual negotiating strength—and the latter depends not on some generalized capacity but the nature of relationships (e.g., degree of mutual interdependence) between groups. While this account sets questions of exchange and power at the heart of structural elaboration, Archer focuses mainly on how different interests can shape the balance between stability and change and the direction taken by the latter.

The consequences of interaction are conceived largely in terms of structural elaboration and its implications for the next evolutionary cycle. This may be acceptable insofar as we are concerned only with process, but it by no means exhausts our concern with the consequences of interaction. I noted earlier that both capitalist economies and welfare states may have consequences for the realization of social roles and the satisfaction of human needs. One need only think of the possible impact of global competition on limiting both employment and taxation. Since people can be conceived as persons and actors as well as agents, we have to consider consequences not only in relation to interests but also in terms of needs and roles. Marx recognized this, and his expectations of socialist transition were based on immiseration (failure to meet needs) and alienation (failure to fulfill roles) as well as the emergence of class interests. Conditions that influence future interaction may include not only relations of exchange and power, but also the (un-)fulfillment of needs and roles.

A focus on explaining stability or change leaves open the question with which we began—of how we can identify consequences as a particular stage in an evolutionary cycle. Archer suggests that the end-point of a particular cycle can be identified when a new configuration emerges that “signals a completely different conditional influence upon subsequent interaction” (Archer, 1995, p.328). She offers as an example the emergence of state educational systems—“representing important transformations of institutional relations which in turn condition future interaction and further educational change” (Archer, 1995, p.337). Thus, end-points represent the culmination of change in a new configuration of conditions.

We might add as further examples,the transition from feudalism to capitalism or the emergence of modern welfare states. These emergent configurations have certainly changed the conditions under which interaction occurs, for example, in such central matters as meeting basic human needs or realizing satisfactory social roles as well as changing the structure of interest groups and relations of exchange and power. However, these examples also raise some further questions. Take, for example, the emergence of a capitalist world economy. Reference to the globalization of the world economy is currently the height of intellectual fashion, but there is little consensus over what this term may mean or how globalization might be measured. If we cannot agree on how to define a new configuration, then we can expect problems in identifying its emergence as the end-point of an evolutionary process. As Wallerstein (1991) notes with reference to this example, "temporal boundaries are by no means self-evident":

> I for example have argued in my writings that the capitalist world-economy came into existence in the long sixteenth century. Others would date it later. A few would date it earlier. And of course not everyone agrees that such a historical system has ever existed at all. Furthermore, it is clear that no one during the long sixteenth century (or virtually no one) yet conceived of this historical system as a system. Indeed, it was not until sometime in the nineteenth century that anyone began seriously to analyze this historical system at all. And it is only in the past 20 years that the concept, the "capitalist world-economy," has taken root in world scholarship, and even now only among some scholars (Wallerstein, 1991, p.142).

Much the same point could be applied to the emergence of welfare states, where similar disputes have developed over just what has emerged—and when. Even if we could answer these questions through various forms of comparative analysis (Janoski & Hicks, 1994), we are still left with the problem of identifying which internal relations and emergent properties can be regarded as constitutive of welfare states. In short, the end-points of evolutionary processes are not given, but have themselves to be theorized in terms of their particular properties and relations.

IN CONCLUSION

In grounded theory, process is identified in terms of successive sequences of interaction, which may be recognized as stages or analyzed in terms of the coding paradigm of conditions–interaction–consequences. However, the terms of discussion of process set out in grounded theory raise some issues about how the analysis of process can be pursued. Some of the main points we have considered in this chapter are summarized briefly below.

First, we need to consider how to incorporate a temporal dimension into analysis rather than rely on a cross-sectional comparison of different slices of time. Here I considered at some length Archer's effort to analyze process in terms of temporal sequence, starting with one configuration of structural conditions and evolving through the interplay of structure and agency into a new configuration.

Second, we need to consider more fully the relationship between structure and interaction. Here I followed Archer's argument that we need to separate structure from agency, conceiving of process as an interplay between the two rather than as mutually constitutive through a sequence of interactions.

Finally, we can develop clearer directions for the analysis of process than those provided by the all-encompassing conditional matrix or the selective tracing of conditional paths. Here I discussed that a range of complementary approaches, including the distinctions between material, cultural and agential forces; the differentiation between persons with needs, actors with roles, and agents with interests; the constraining role of existing vested interests and opportunity structures, and the explanatory promise of analyzing situational logics.

Overall, the picture of process that emerges from Archer's account may be considered much more complex than presented in grounded theory, but also more focused and amenable to analysis. For one thing, process is analyzed as evolution over time, rather than as a series of slices of time viewed in cross section. Moreover, we have some guidance as to possible paths of process, expressed in the articulation of (in-)compatibilities in structure, the differentiation of situational logics, the distinction between various forms of agency or interaction, and the analysis of relations of exchange and power. These signposts (while not exhaustive) at least give some conceptual direction to the identification of key conditions that we lack in grounded theory—where the conditional matrix potentially encompasses all conditions, while any selective focus on tracing conditional paths seems essentially arbitrary.

Though my discussion has been rather critical, and, no doubt (despite its length), far from exhaustive, I hope it has also been constructive in identifying some potentially useful conceptual tools for the analysis of process.

CHAPTER 9

Making Fit Theory

In this chapter, we consider what kind of guidelines grounded theory offers for generating theory, notably the distinction between substantive and formal theory. We consider what, if anything, is distinctive in grounded theory about its approach to theory. In this respect, I suggest that grounded theory offers a method of theory generation that mixes two modes of theorizing that are traditionally counterposed: the idiographic and nomothetic perspectives. We consider whether this mixed method of theorizing can be justified and what its methodological implications might be. We also consider the criticisms which Glaser and Strauss direct against other modes of theorizing; criticisms intended to situate the grounded theory approach as well as point up the weaknesses in alternative approaches. This chapter therefore engages with the question of what constitutes a "fit" (i.e., productive) theory in social research. We leave the question of ensuring that theory fits the objects of our inquiries until the next chapter.

THEORIZING IN GROUNDED THEORY

In the interests of clarity, we need first of all to consider what is meant by "theory" in grounded theory. A useful starting point (but only a starting point) is the distinction Glaser and Strauss draw between substantive and formal theory. In the original text, the distinction is drawn in the following terms:

> By substantive theory we mean that developed for a substantive, or empirical area of sociological inquiry, such as patient care, race relations, professional education, delinquency, or research organizations. By formal theory, we mean that developed for a formal, or conceptual, area of sociological inquiry, such as stigma, deviant behavior, formal organization, socialization, status congruency, authority and power, reward systems or social mobility (Glaser & Strauss, 1967, p.32).

Glaser and Strauss try to distinguish here between empirical and conceptual areas of inquiry, though, as we see, the distinction is rather hard to grasp.

Take, for example, the contrast they offer between delinquency (which is cited as an empirical area of inquiry) and deviant behavior (which is cited as a conceptual area of inquiry). On what criteria are these assignations made? It is certainly not obvious why delinquency should be defined in one way and deviant behavior in another. Perhaps delinquency represents a form of deviant behavior in a specific context—that is, the deviant behavior of young people? Glaser does develop an argument along these lines in a further example, this time of the organizational careers of scientists. Here, he suggests that a study of scientists careers is substantive while a study of organizational careers is formal. Perhaps, then, delinquency is defined as a substantive area because it is refers to deviance among young people, just as scientific careers is seen as a substantive area because it refers to scientists?

Strauss and Corbin do seem to draw the distinction in just these terms, suggesting that:

> . . . any substantive theory evolves from the study of a phenomenon in one particular situational context. A formal theory, on the other hand, emerges from a study of a phenomenon examined under many different types of situations (Strauss & Corbin, 1995, p.174).

Thus behavior studied in one context (juvenile delinquency) generates substantive theory but if studied across many different situations (deviant behavior) it generates formal theory.

One difficulty with this distinction lies partly, I think, in the inconsistency between the principle that is proposed and the examples that are offered. For though substantive theory is supposed to study "a phenomenon in a particular situational context," we may doubt whether that context is sufficiently specified in the examples provided. We can regard "scientific careers" as a subcategory of "organizational careers" but this does not in itself make the area substantive rather than formal. So too with "juvenile delinquency" and "deviant behavior." This point is readily demonstrated if we consider other taxonomic levels.

For example, we might distinguish among different sorts of scientific careers—in different subjects, for example (physicists, chemists, geologists, etc.); in different organizational settings (academic, industry, research institutes, etc.); or in different forms of inquiry (theoretical, experimental, observational, etc.). Even if we combine these to identify as the subject of our study (for example) experimental physicists working in industrial laboratories, we have not sufficiently specified "a particular situational context," but only a further subdivision of a (formal) field of inquiry. The converse also applies, for we could take organizational careers as a subdivision of careers—but this, of course, does not somehow transform organizational careers from a formal into a substantive topic.

Distinctions Through Space and Time

A more meaningful distinction between substantive and formal theory might be drawn in terms of temporal and spatial characteristics. The specification of "a particular situational context" usually involves its location at a particular time and place. Glaser and Strauss (1967, p.81) recognize as much—though only in passing—when they complain that most formal theory is produced by researchers who "have not escaped the time and place of their substantive research" but simply abstracted from it. The nub of their argument is that formal theory should be produced through the hard work of comparative analyses—and not simply by notching up the level of (taxonomic) generality by dropping reference to the empirical context in which the research was done. This seems in part a (veiled) criticism of the single case study as a method of generalizing or generating theory—a point to which I return.

Meantime, we should note that "time and place" are largely presented by Glaser and Strauss as restrictions from which to escape through the logic of constant comparison. Even substantive theory "evolves" from a particular situation context through a comparative approach—for example, comparing scientific careers with those in the law or the military (Glaser, 1978, p.144). Indeed, it is argued that "the process of comparative analysis is the same for generating either substantive or formal theory"—though the latter does require a wider range of research at higher level of abstraction (Glaser & Strauss, 1967, p.82).

If substantive and formal theories are generated through the same method, differing only in levels of generality, it is little wonder that we have difficulties in sustaining the distinction between them. Once we take account of time and place, however, we can distinguish between them on a surer footing. The reason is that these are variables that specify a phenomenon uniquely: no other event (we can safely assume) can occupy the same time and place. This is so even though people differ in their experience and interpretation of it—it is not the event which is rendered "multiple" but merely our interpretations of it. The "particularity" of a situational context, therefore, can be defined in terms of temporal and spatial dimensions: we can specify "experimental physicists working in industrial laboratories" by reference, for example, to experimental physicists working in Bell Laboratories during the 1980s.

From this perspective, a substantive theory is not designed to explain phenomena at a lower level of generality; rather, it is designed to account for a particular phenomenon where that particularity is defined in terms of time and space. Thus a substantive theory has a particular subject (specified in time and space) while formal theories have general subjects, which at least to some extent escape these spatial and temporal boundaries. Thus we can study the evolution of the British welfare state in the postwar years (a substantive theory)—or of welfare states in general (a formal theory).

Note that a substantive theory concerned with a particular subject can be narrow or broad, depending upon one's research interests. Take, for example, the American War of Independence. To explain this event (or more accurately, series of events) through a substantive theory, we might have to take account of the configuration of structural and cultural conditions in both the New World and the Old World and explicate the process that culminated in revolution and ultimately in independence. We could also consider the ramifications of the American Revolution—which continue to affect us even today. There is no reason, therefore, to equate a substantive focus on particular events with a narrow concern with insignificant incidents that have no wider implications for our own situation. Much of our contemporary world is shaped by particular events—including several other revolutions, in France, Russia, and China but also industrial and technological—that have had a profound significance for how it has evolved.

Theorizing Conjunctures and Generalities

At first sight, the focus of substantive theory on particular events may seem to require a methodology distinctly at odds with that associated with formal theory. This requirement for a distinctive methodology seems to stem mainly from the need in substantive theory to locate events uniquely in terms of space and time—in other words, to delineate the (spatial and temporal) context in which the phenomenon occurs. Because it locates events in a specific context, substantive theory can take account of the complexities of particular "conjunctures."

The dictionary defines a conjuncture as a "combination of events" or "state of affairs"—deriving from "jungere" meaning "join" and "com" meaning "together." Following Archer's account of the interplay of structure and agency, we can use the term "conjuncture" to refer to the joining together of structure (states of affairs) and agency (events) at any particular point in time. This joining together suggests another aspect of substantive theory: its holistic power. The focus on a particular phenomenon invites (though it does not require) a holistic approach, that aims always to grasp the phenomenon as a whole rather than through some aggregative summation of its various parts. Instead of trying to identify a series of variables that act independently, a holistic perspective considers variables in terms of how they unite, intersect, or otherwise relate (for example, through "internal and necessary" relations) within the context as a whole.

In contrast, formal theory tends to abstract from any particular context, offering general explanations that account for events regardless (apparently) of their particular space-time location. These explanations tend to be couched in terms of key variables—using the methods of agreement or difference discussed earlier, for example, to identify constant conjunction across a range of cases and so provide a basis for causal inference. Where a large number of cases is included, statistical analysis may be employed to assist in the identification of patterns of covariation. Variables are assumed to be constant over time and space rather than evolving in the light of changing circumstances (or we could not compare like with like). Insofar as it abstracts from limitations of place and time, formal theory tends to be ahistorical and aholistic. In other words, it does not try to explain events in terms of particular configurations and processes grasped as a whole. Ideas tend to be connected through propositional logic (if A then B) rather than configured as a narrative (a story of how events unfold).

For example, take the evolution of welfare states. The origin of the welfare state in Britain is sometimes located in the liberal era preceding the First World War. Explanations of liberal prewar legislation (including the first tentative forays by the state into insurance for health and unemployment) are couched in terms of a complex interplay of factors, including Liberal Party fears about its political future given the rise of the Labour Party; imperial ambitions sharpened by the growing competition from Germany, which itself provided an example of state involvement in welfare; the civil unrest and conflicts precipitated by the emergence of mass unemployment; a gradual reassessment of the role of the state, encouraged by the growth of the civil service; the problems posed by inadequate forms of insurance; growing evidence (partly based on new survey techniques) of severe forms of poverty and malnutrition; and the political vision and administrative drive of such key figures as Lloyd George, Churchill, Beveridge, and the Webbs. Different authors may emphasize different factors, but all would agree that the "Liberal Reforms" (as they became known) originated in a particular conjuncture, marked by the rise of foreign competition, internal class politics and so forth.

By contrast, formal theory seeks explanations of the evolution of welfare states in terms of key variables, such as the growing power of social democratic parties or labor movements, the growth of economic capacity and state spending, changing levels of employment, and unemployment and the like. For example, we may associate higher welfare expenditures with situations where social democracy is strong and unemployment is high but economic capacity is growing. We can add in other variables, such as the degree of industrialization and urbanization, demographic change (such as an aging population), the degree of democratization and even involvement in international institutions promoting social legislation (Usui, 1994, pp.260–1). The aim is to identify and isolate those (causal) factors that can "explain" most of the variation between countries in the degree to which the state intervenes in welfare.

INTENSIVE AND EXTENSIVE THEORIZING

As we have seen, Glaser and Strauss suggest that though they vary in terms of generality, both substantive and formal theorizing in grounded theory involve much the same methodology. This is an especially notable characteristic of grounded theory, as different theoretical modes are usually associated with quite different methods of inquiry.

A useful contrast between different theoretical approaches can be drawn in terms of "intensive" and "extensive" forms of theorizing (Isaac *et al.*, 1994, pp.94–5; Sayer, 1992, p.243). In relation to historical/comparative studies, for example, Isaac and colleagues note a number of contrasts in analytic approach:

Intensive	*Extensive*
Case-centered focus	Variable-centered focus
Qualitative-historical approach	Quantitative-ahistorical approach
Holistic configurations across time	Variables over time within or between cases
Internal process/event sequence of substantial relations	External quantitative relations often reduced to summary singular point emphasis
Changing qualitative outcome	Changing quantitative outcome
Complex conjunctural determination	Linear, additive "effects" of single variable or sum of variables
Temporally heterogeneous path/sequence dependent with possible multiple paths to single outcome	Temporally homogeneous with event sequence and path conditions typically ignored
Deviant cases centrally theorized	Deviant cases eliminated or stabilized for fit
Close narrative	Wider perspective

Source: adapted from Isaac *et al.* (1994, p.94)

Each mode of theorizing can be seen as limited in some ways.

The restrictions associated with intensive theorizing stem from its focus on particular events. As I suggested earlier, this is not because intensive theorizing must be confined to local scale. Indeed, Tudge (1996) argues that a proper perspective on human history can only be achieved if we consider it as a whole—a period of five million years—the past few thousand years being, from this standpoint, merely "the present!" The restrictions seem not so much of scale as of relevance and application. If there are general conclusions to be drawn by social theory, a focus on explaining particular conjunctures may seem an awkward way of reaching them.

In a search for generalizations, we may turn to extensive theorizing, but this is associated with other pitfalls. We considered earlier some of the difficulties of identifying causality on the basis of invariant patterns. The problem of rendering an intelligible account of how things work on the basis of correlations among variables is exacerbated if we lack intimate knowledge of complex processes which shape events in particular contexts. This problem is especially acute if we accept that events are shaped by people with their own particular perceptions, purposes, and projects.

Thus we seem to be faced with an awkward choice between these different forms of theorizing. However, grounded theory seems to offer a way out of this impasse. Insofar as grounded theory embraces both substantive and formal theories, we might reasonably expect it to straddle the intensive/extensive divide.

GROUNDED THEORY AS A MIXED METHOD

In some respects, theorizing in grounded theory seems to conform to the strategies of the intensive approach. It is associated (rightly or wrongly) with a predominantly qualitative methodology; it emphasizes process; it focuses on qualitative rather than quantitative relationships among variables; it incorporates rather than excludes deviant cases; and it encourages (without insisting upon) a narrative account.

On the other hand, it also conforms to some of the canons of extensive inquiry, notably its focus on variables (rather than cases); its use of comparative analysis to identify causation; its assumption (perhaps unwarranted) of homogeneity with regard to variables; and its ambition to produce formal theory freed from constraints of time and place.

This straddling of a methodological divide is undoubtedly one of the great attractions of grounded theory, since it appears to offer a "middle-way" approach that escapes from the restrictions of intensive theorizing while avoiding some of the pitfalls of extensive theorizing.

One distinctive strategy that grounded theory offers lies in the application of comparative methods to intensive inquiry. As noted earlier, substantive theory is distinguished from formal only in terms of levels of generality, and not in terms of methods of inquiry. Thus in grounded theory, substantive theory is based on methods more commonly associated with extensive inquiry, most notably an orientation to comparing variables (categories) rather than analyzing cases. The comparative method requires selection on the basis of those categories of theoretical interest, across a wide range of situations that bring out the patterns of similarity and difference between categories. These patterns are analyzed in terms of their qualitative rather than quantitative properties (though as Glaser and Strauss suggest, nothing in principle excludes a numerical approach). Relationships within and between categories are identified through repetition which brings out underlying uniformities:

> The comparison of differences and similarities among groups not only generates categories, but also rather speedily generates generalized relations among them (Glaser & Strauss, 1967, p.39).

These emergent relations can then be "verified as much as possible through the course of research" (Glaser & Strauss, 1967, p.39).

Another distinctive strategy lies in the application of intensive methods to comparative inquiry. This is evident in some of the methods that are advocated to achieve theoretical integration by connecting ideas. For example, Strauss and Corbin suggest that "to achieve integration, it is necessary first to formulate and commit yourself to a story line" (1990, p.119). A "story" is defined as "a descriptive narrative about the central phenomenon" while a "story line" is "the conceptualization of the story"—which is equated with the selection and elaboration of "the core category" (Strauss & Corbin, 1990, p.116). Here the idea of developing a narrative account is applied to the task of articulating key relationships between concepts. This implies of course that these relationships can conform to a narrative structure.

There may be various kinds of narrative structure (just as there are different kinds of stories) and Strauss and Corbin do not make explicit quite what they have in mind. The suggestions they do make point in rather different directions. On the one hand, they suggest that integration involves axial coding at a higher level of abstraction. This implies that the coding paradigm of conditions—interaction—consequences can provide the necessary narrative structure—just as a story can progress from its start, through a sequence of events, to a conclusion. Here the explication of a story line hints at a holistic, conjunctural, and multifactorial account of how events unfold.

On the other hand, Strauss and Corbin also suggest that identifying the story involves selecting and naming the main phenomenon (or core category) around which the research revolves. In other words, it offers a condensed account of the key relationships and concepts utilized in the research. Here the explication of a story line seems to involve selecting and relating key variables—apparently without any particular reference to how events unfold over time.

Another method of integration in grounded theory involves "theoretical sorting." What gets sorted are the theoretical memos that the researcher has recorded in the course of the project. Glaser asserts that this is an essential step as it "begins to put the fractured data together again":

> If the analyst does omit sorting, he will indeed have somewhat of a theory, but it will be linear, thin and less than fully integrated. It will not be the rich, multi-relation, multi-variate theory that can be generated by sorting. The theory may have overall integration, but it will lack the internal integration of connections among a great many categories (Glaser, 1990, p.116).

Thus theoretical sorting is needed to produce "a dense, complex theory" (Glaser, 1978, p.117), though there are no specific analytic (as opposed to practical) rules as to how this should be done. As usual, no ready-made schemes should be adopted (though various theoretical codes may be indicated) but instead an outline should be allowed to emerge:

> [The analyst] should simply start sorting the categories and properties in his memos by similarities, connections and conceptual orderings. This forces patterns which become the outline. . . In sorting, the analyst is constantly moving back and forth between memos and a potential outline, working with it so everything fits (Glaser, 1978, p.117).

How such "connections" are made, how an outline is identified, and how "fit" with the outline is achieved remains rather a mystery—a matter of "catching on" to how to make such "complex collations" (Glaser, 1978, p.117). However, the emphasis on developing a dense and complex theory does introduce theoretical ambitions more often associated with intensive inquiry into the domain of comparative analysis. Of course, a price perhaps has to be paid for this, in the need to select a core variable in order to keep this density and complexity within manageable limits.

There may be other costs, too. Although intensive methods may be expensive in terms of time and resources, that is because they depend on an intimate knowledge of the context and evolution of events, including the various perspectives of different actors. As I suggested earlier, it is not obvious that this knowledge can be obtained through the "smash and grab" techniques epitomized by the quick afternoon call. In grounded theory this is legitimated by a focus on variables rather than cases—but this tends to undermine any claim to "ground" theory in a thorough examination of particular cases. What is compared in grounded theory is data relevant only to selected variables, abstracted from context (through theoretical sampling) even at the point at which much of the data are collected.

Another cost may lie in its dilution of comparative method, as the logical requirements of the methods of agreement and difference are loosened and the associated methodological rigor reduced. The most striking manifestation of this can be found in the treatment in grounded theory of negative cases. In inductive logic, a negative case has to be dealt with either by redefining the nature of the phenomenon under investigation (so that it no longer constitutes a case at all), or by rejecting the hypothesized relationship outright. In grounded theory the relationship that survives as the negative instance is simply taken as an example of further variation. The addition of new conditions (if they can be identified) certainly adds to the complexity of the theory—but there is surely something rather disconcerting about such ad hoc theorizing that so resolutely refuses any refutation. It may be appropriate in the context of intensive theory, where we only need to explain one phenomenon in particular. But is it so acceptable in the context of extensive theory, where we want to generalize about a range of phenomena?

> Some Possible Costs of Mixing Intensive and Extensive Methods of Inquiry
>
> - *Inquiry has to be made manageable by focusing on a core category that may be selected arbitrarily*
> - *Data acquired in the later stages of inquiry may be inadequately grounded*
> - *The comparative method cannot be applied rigorously to multiple and conjunctural forms of causation*

OTHER MODES OF THEORIZING

Still, grounded theory at least attempts to develop a hybrid methodology and, by offering a middle way, offers to free us from what may seem an invidious choice between intensive and extensive approaches. In this, Glaser and Strauss show a more eclectic approach (in my view, much to their credit) than many of their social science colleagues, who seem to regard various modes of theorizing as forms of faith rather than instruments of inquiry.

I noted earlier that intensive methods are discounted by some as "unscientific" while extensive methods are dismissed by others as "positivist." If one of the great virtues of social science is to teach tolerance of other points of view, this is sadly not a virtue much in evidence among its practitioners! Given the great difficulties confronting any form of social research, it is striking how often researchers claim that their approach is the only way it can be done. Thus Denzin and Lincoln (1995, p.5), for example, bemoan the "long shadows" cast over qualitative research by the "positivist" and "postpositivist" traditions—citing grounded theory as a (lamentable?) example of the latter.

Typical of this dismissive stance is the following comment by Wallerstein on the spatial and temporal assumptions of "idiographic" and "nomothetic" modes of theorizing:

> The TimeSpace of our nomothetic social scientists seems an irrelevant illusion. The TimeSpace of our idiographic historians—events in immediate geopolitical space—seems a series of self-interested inventions about which there will never be agreement as long as political discord exists in the world. In neither case is Time-

> Space taken seriously as a basic ingredient of our geohistorical world (Wallerstein, 1991, p.144).

By "idiographic" methods, Wallerstein is referring to a historical perspective that focuses on particular events defined uniquely in space and time—giving us "a chronology and thereby a narrative, a story, a history that is unique and explicable only in its own terms" (Wallerstein, 1991, p.137). This is close enough to the way I have presented intensive inquiry—so we can think of intensive methods as a way of pursuing idiographic theorizing. By "nomothetic" theorizing Wallerstein is referring to efforts to generalize "across time and space"—that are based on extensive methods of inquiry. So this ready dismissal of idiographic and nomothetic methods as deficient in their analysis of "TimeSpace" is very relevant to the virtues of seeking a middle way. But does a middle way (assuming we can find one) have to become, as Wallerstein implies, the way of developing social theory?

Let us look at Wallerstein's arguments. His attack is launched against two "nominally antithetical" positions—the idiographic and the nomothetic—which he suggests "have dominated social thought since at least the middle of the nineteenth century" (Wallerstein, 1991, p.137).

> Idiographic: *Concerned with the individual, pertaining to or descriptive of single and unique facts and processes.*
> Nomothetic: *Of or pertaining to the discovery of general laws.*

The idiographers have focused on unique events, defined in terms of their occurrence at a particular place and time. Against such a superficial view of events, that takes them at face value, Wallerstein points out that the very idea of an "event" is a social construction. Some events are recorded as such at the time; some are recorded by historians at a later time; and most are never recorded at all. What is discerned as an event worthy of recording depends upon its significance—but even of those which seem significant Wallerstein (1991, p.137) asks "do these in any sense matter"? Citing Braudel, he argues that, seen in a longer-term perspective, "events are dust." In this longer-term perspective, it is not ephemeral events but structural change (such as the growth of a capitalist world economy) and cyclical patterns in development (such as economic expansion and contraction) that are significant.

The nomothetic position, by contrast, takes a perspective that is too long-term even for Wallerstein—indeed, it eliminates altogether considerations of space and time from theorizing about society. It involves a search for "universal eternal patterns of human behavior." Thus Wallerstein castigates nomothetic theorists for ignoring the realities of social change—slow though that change may be, when considered in terms of structural evolution.

Wallerstein (1991, p.138) argues that these apparently antithetical views in fact share a common position. They are "but two modes of trying to escape from the constraints of historical reality"—the one through a narrow focus on particular events and the other through a general focus on the universal. Following Braudel, he argues for a middle way that recognizes different times between the instant and the eternal. He has two particular types of time in mind. These are structural time (which refers to the slow evolution of enduring structures over centuries) and cyclical time (which refers to the repetitive patterns created by oscillations in structural functions). To theorize, then, requires a historical perspective, but one that theorizes about long-term social structures, their repetitive cycles, and their slow transformation over time.

Although Wallerstein's argument may be a vital corrective against antihistorical perspectives, the case he puts seems overstated. Even if we accept the main point of the argument—that events occur through time, which is neither instantaneous nor eternal—it does not follow that we should analyze events only in terms of structural or cyclical time. It may certainly be very valuable to recognize these temporal perspectives—but only, I suggest, as two options among others. If we acknowledge that there can be a multiplicity of temporal perspectives, then we need to make these temporal perspectives not only explicit, but also appropriate. Among other things, this depends on the balance of structural reproduction and transformation, and also on the direction and pace of change—with circular or cyclical patterns as only one option. We cannot simply assume with Wallerstein that "events are dust" as though all events were of equal status.

If we adopt this more flexible approach, then it seems less reasonable to discount so emphatically the contributions of the idiographic and nomothetic positions. Neither may be quite so "ahistorical" as Wallerstein suggests. Events can be large-scale as well as small-scale, and although an idiographic approach may focus on a particular event, it is also an event with a past and a future. In other words, we ask not only what happened, but how (or why) it happened and what happened next (as a result). With these questions, we can travel far and wide across both temporal and spatial dimensions. This is so, even if the event in question can be as narrowly defined in time and space as the moment that Rodrigo de Triana first saw land and Columbus (falsely claiming the credit) "discovered" America. As John Muir (1996) wisely wrote, "when we try to pick out anything by itself, we find it hitched to everything else in the universe." Wallerstein may be correct to argue that events are socially constructed and as such subject to political interpretation. No doubt the native inhabitants of the New World did not think of themselves as "discovered" by the natives of the Old World. But such is the condition of social science—whatever the time scale.

Indeed, the social construction of events undermines any claim that idiographic theorizing can focus on any particular event in historical isolation. It is only in a wider historical context that the event can be constituted as such. If the American Revolution was the only revolution ever to have occurred in history, we would not call it the American Revolution. We define the event in terms of knowledge derived from a wider context. My point is clearer, sadly, if we consider wars, which are so frequent that they can even acquire numbers as well as names. Thus the Second World War can only be understood as "a war" because we bring to this event the knowledge derived from other conflicts. Even the recognition and recording of an event reflects a wider appraisal that imbues the particular phenomenon with the required significance to be recognized and recorded as such. When we try to identify and explain a particular phenomenon, we do not suddenly bury our heads in an ahistorical sand that obscures the rest of our knowledge from view.

The nomothetic approach, on the other hand, is dismissed as ahistorical because it aims at universal, eternal generalizations about the human condition; the aim, as Glaser and Strauss put it, is to "escape" from the constraints of time and place. But it need not escape from considerations of time and place altogether, but only of a particular time and place. To generalize beyond one population, we do not need to generalize to all populations. To generalize beyond one location, we do not need to generalize to all locations.

Natural science may produce generalizations that are truly atemporal. For example, Newton's equations are indifferent with regard to time—his laws (generalizations) hold whether time is running backward or forward. Gödel proved that Einstein's equations allow travel backward in time (though only in theory, of course). But a moment's reflection can confirm that no such atemporal generalization is possible in social science, which is located in and concerned with evolutionary and historical time. However, this does not exclude the possibility of generalization.

When billiard balls kiss, they obey universal laws that describe their interaction. When human being kiss, they obey no such laws. But that does not mean that we do not or cannot generalize about kissing. Indeed, unless we do generalize about kissing, we are liable to find ourselves in deep trouble if and when we try to practice the art—for we will not know when to kiss, whom to kiss, or how to kiss them. When is a peck on the cheek called for (the visiting aunt or uncle) and with whom is a more passionate kiss permitted? (I leave that to your imagination.) To answer such questions, we need to be able to generalize about what behavior is appropriate, when, and with whom. But while these generalizations may escape the constraints of a particular time and place (it is not any particular kiss we need to know about) they are not unbounded by considerations of space and time. There is always a social context, both temporal and spatial, within which generalizations may work (though we all make mistakes); but beyond which these generalizations simply do not apply (though we may learn this the hard way). Of course, in the absence of time travel, we are more likely to encounter spatial than temporal boundaries to our generalizations. If and when we visit a different community or society, we know that we may have to learn very quickly a quite new set of rules.

This implies that generalization about society or social interaction must always be bounded by space and time, even if these boundaries are often allowed to remain implicit. There is always a (tacit) context within which extensive theory is assumed to apply. Thus generalizations apply neither to the particular nor to the eternal, but to events within some (implicitly) bounded space and time in which they are assumed to occur.

It is sometimes assumed that theory equates with generalizations. This view tends to arise from a misleading impression that the aim of science is to produce "laws." A scientific law is taken to refer to universal regularities, such as the variation of temperature of a gas with changes in pressure and volume. But there are several reasons why this equation (of science with law-like generalizations) is unfortunate. For a start, it gives an inadequate account of science. Natural scientists are often seen as privileged because experiments are ideally suited to examining how things work by allowing manipulation of a variable (in controlled conditions) and then observing the result. Social scientists are then cast in the role of poor relations, emulating the scientific process as best they can by devising various forms of (unsatisfactory) "pseudo-experiment." It is conveniently forgotten how often natural science progresses through observation rather than experiment (think of one of the most successful theories science has produced—that of biological evolution). And in one respect, the social sciences have a significant advantage over the natural sciences—for as well as observing, they can communicate with the subjects that they study.

Scientific law: correct statement of invariable sequence between specified conditions and specified phenomenon, regularity in nature.

In any case, natural science is by no means focused on producing such law-like generalizations. For these laws only describe behavior, rather than explain it. Summarizing the regular relationship between volume, pressure, and temperature in a law does not explain why gases behave as they do. Moreover, these laws can operate as calculating devices in closed systems (knowing the pressure and volume, you can calculate the temperature), but only so long as nothing else changes. Insofar as society is an open system (as indeed are many physical processes) it does not begin to approximate the conditions that would make such calculations possible (Sayer, 1992, pp.125–130). In a neat analogy, Archer (1995, p.70) suggests that trying to close a social system by controlling variables is akin to closing the stable door on a horse that knows how to open it. Finally, much scientific effort is directed not to identifying regularities or producing laws but to understanding and explaining particular events—such as the origin and evolution of the universe or of life on earth.

A Rapprochement

These remarks (if reasonable) permit and perhaps encourage a rapprochement of sorts between idiographic and nomothetic theorizing. On the one hand, the idiographic position cannot be sustained without support from our general knowledge—about revolutions, welfare states, hospital wards or whatever. We can study the particular, but in doing so, we cannot use only knowledge germane to the particular—we must call upon general knowledge that aids us in identifying particular events, putting them in context, recognizing the processes that are at work, and how these produce particular results. The particular configuration may be unique, but many of the various elements that constitute that unique constellation need not and indeed cannot be. On the other hand, the nomothetic position cannot be sustained without some specific context, albeit implicit, within which its generalizations can be applied. We may generalize about revolutions, welfare states, or hospital wards, but our generalizations only make sense within political and social contexts in which these phenomena are possible. Such a rapprochement may in itself suggest a middle way for social theory, since, instead of two antithetical positions, we are left with differences of focus and emphasis (Fig. 9.1).

In idiographic theorizing, we focus on the particular and emphasize specif-

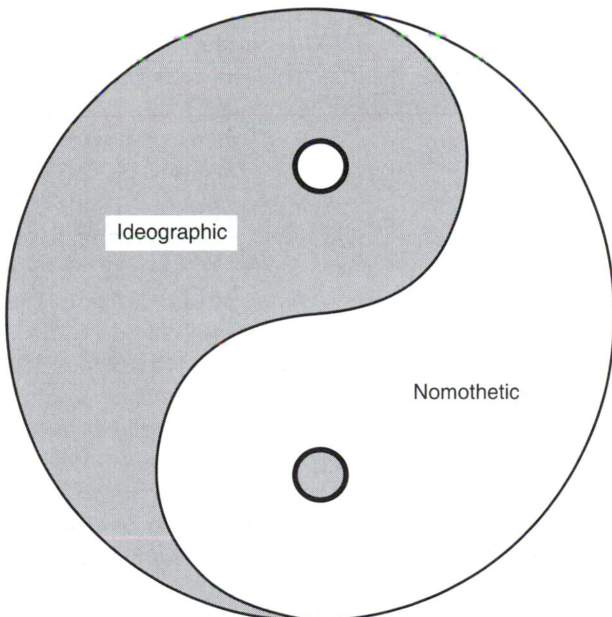

FIGURE 9.1 Idiographic and nomothetic theorizing as mutually implicated.

ic complex configurations that nevertheless can only be understood because we bring our general knowledge to bear upon them. In nomothetic theorizing, we focus on events abstracted from their particular time and place, emphasizing what they may have in common, but only within the boundaries of a given spatial and temporal context. As I suggested earlier, both approaches seem eminently reasonable. We could not survive in the world without understanding particular events in all their complexity; but nor we could not survive without comprehending some generalizations about how and why things work out as they do. And each approach depends to some extent on the other.

Criticisms of the Nomothetic Approach

While this rapprochement lends support to the use of mixed methods in grounded theory it also suggests that the criticisms of alternative modes of theorizing that inspired grounded theory themselves need to be reappraised. The claims of grounded theory to offer a middle way are partly based on a rejection of alternatives. While there is little point in rehearsing battles fought long since, there may be some merit in questioning the terms of the debate insofar as these criticisms situate the claim of grounded theory to a distinctive mode of theorizing, and inform the methods that Glaser and Strauss regard as appropriate for that mode.

The most obvious target for criticism, from a grounded theory perspective, is the nomothetic approach, as embodied in the kind of armchair theorizing about society which Glaser and Strauss deride. They contrast grounded theory with "theory generated by logical deduction from a priori assumptions"—assumptions that are "ungrounded" and can therefore "lead their followers far astray" (Glaser & Strauss, 1967, pp.3–4). There are two targets of criticism here: the "ungrounded" and "a priori" assumptions and the "logical deductions" made from them. Glaser and Strauss do not distinguish these targets, preferring to reject both. They do not, therefore, consider the merits of logical deduction from "grounded" premises as a mode of theorizing. This leads them into some difficulty, when it comes to the question of how to theorize in areas of enquiry where some grounded theory has already been done. Do we have to start from scratch or can we build on the results of previous inquiries? We cannot accept the premises supplied by existing grounded theory as a starting point unless we also accept logical deduction from these premises.

Glaser and Strauss are also generally hostile to the use of logical deduction as a method of pursuing inquiry within a project. They caution continually against speculative forms of theorizing that are unsupported by data. Not only is logical deduction dismissed as speculation (though logical conclusions from established premises are hardly speculative), but they invariably see such speculation as an invitation to "force" the data to fit the theory. Thus they argue that the properties of a category must be derived from observation, not deduction. Although later, Strauss and Corbin are rather less critical of deductive thinking, even acknowledging that it is "very much a part of the analytic process" (1990, p.148), it figures more as an emergency stop-gap (to be called upon when all else fails) than as a powerful mode of theorizing in its own right.

As I argued earlier, the argument, that deductive thinking "forces" data, confuses two different targets—logical deduction, on the one hand, and whether data are forced to fit the categories so deduced, on the other. There is no need to force data to fit categories, and indeed the rhetoric (if not always the practice) of this mode of theorizing is in fact to look for data that do not fit. This is because the method proceeds by testing deductive conclusions against evidence with the implication that the evidence may or may not fit. If it does not, then the premises from which the conclusions were drawn must be rejected (at least as they stand). Though Glaser and Strauss (1967, pp.261–2) note "several kinds of license to preserve speculative theory in the face of contrary evidence" this does not invalid the procedure but only its misapplication. Properly applied, logical deduction can support a "dialogue" with data in which the deductive conclusions are continually checked against the available evidence. Moreover, sometimes it may be as important to theorize about what is "missing" from the data as it is to theorize about what is there. This kind of dialogue, which allows some distance to develop between ideas and data, is not acceptable in grounded theory because of the insistence that all ideas must fit.

Suppose we want to study how physical contact develops between patients and staff in hospital wards. We find that there are various ways in which we ca have physical contact, some more "supportive," some more "instrumental," and some more "forceful" than others. We may observe that patients develop various patterns of contact with different categories of staff—such as auxiliary and nurses. Now suppose we observe that auxiliaries and nurses both tend to develop more supportive forms of contact with patients with whom they interact on a regular basis. Presented in logical form, we might claim that "if staff have regular interaction with patients, then they tend to develop more supportive forms of physical contact with patients." From this premise, we can (logically) derive a hypothesis: doctors (also staff) who interact with patients on a regular basis also develop more supportive forms of contact. Note that this hypothesis is not derived from observation—we have yet to observe what doctors do. In grounded theory it appears that this procedure would be regarded as forcing data to fit a preconceived question—in this case, about the nature of the physical contact doctors have with patients. It is fine if this observation emerges from the data, but invalid to look for it in the data. But note that in looking for it in the data, we are not looking for a preconceived conclusion, but only for the (unpredictable) answer to a preconceived question.

We have made the data fit by directing our observations—but we have made the data fit the question, and not a preconceived result.

Ironically, this may approximate the procedure that underlies theoretical sampling, which seems to involve the selection of groups or situations that allow the further exploration of properties already identified in the course of the research. For example, Glaser and Strauss (1967, p.55) suggest that groups may be chosen to increase "the possibility that the researcher will collect much similar data on a given category" and also "a few important differences." Thus we could select our doctors for study given our interest in the category "staff physical contact with patients"—but will "similar data on a given category" emerge (or even "a few important differences") unless we make observations which are relevant for this purpose? If this does involve deductive thinking, then it might be better to make it explicit in order both to sharpen the questions to be resolved and the evidence required to resolve them. But as Glaser and Strauss envisage a more focused form of interviewing as the research proceeds, perhaps this is exactly what they have in mind?

Logical deduction itself is not speculative, but the premises employed in deductions may be so. Such speculative premises form the other target of criticism by Glaser and Strauss. They disparage the "grand theory" of the day as failing to connect with empirical realities. By contrast, grounded theory—whether substantive or formal—is presented as theory of the "middle" range. While formal theory may escape the confines of time and place, to remain grounded it has to be generated through or interpreted by substantive theories. Again there may be some confusion, here, over what exactly is intended as the object of criticism. On the one hand, we have tendency of sociologists to try to generalize on a universal scale without sufficient regard to the spatial or temporal boundaries within which any generalizations must apply. On the other hand, we have speculative theory, of which such grand generalizations offer only one example. Speculation in science need not (and arguably does not) involve such flights of fancy, for typically the imagination is disciplined by what has already been established.

So, at least, claimed Richard Feynman, surely one of the most speculative thinkers of the twentieth century. For Feynman, scientific creativity was not indulgence in fanciful speculation but "imagination in a straitjacket" (Gleick, 1994, p.324). According to one of his biographers, Feynman's methods were once described by fellow physicist Gell-Mann as follows: "You write down the problem. You think very hard. . . Then you write down the answer" (Gleick, 1994, p.315). Like Einstein, whose speculation that the universe is curved derived from a thought experiment (what it would be like to ride on a beam of light), Feynman tried to visualize how problems might be solved. But the visualization was always disciplined by evidence: "whatever we are allowed to imagine in science must be consistent with everything else we know . . ." (cited in Gleick, 1994, p.324). I am not suggesting that we can emulate the genius of a Feynman or Einstein, but simply noting that speculative theorizing (suitably disciplined) played a crucial role in how they did science. Thus speculation can be disciplined by existing knowledge; whereas in grounded theory, it is disciplined by only by the current research project—or banned altogether.

Explanation of the extinction of the dinosaurs some 65 million years ago offers an example of how "asking the right question" can illuminate data. Speculation about the events which led to that extinction included the idea that a large meteor may have been responsible. This speculation led to the discovery of iradium dust (which only comes from meteorites) in deposits dating from 65 million years ago in excavations at various sites around the world. If the question had not been asked, no one would have checked the deposits for iradium dust. This underlines another important point about speculation—that it should produce hypotheses that can be tested against evidence. Although there may be many occasions where this rule seems honored in the breach, that in itself does not undermine the role of speculation in generating theory. Once again, it is important to distinguish commitment to asking a particular question and commitment to producing a particular answer—a distinction that Glaser and Strauss fail to make:

> Potential theoretical sensitivity is lost when the sociologist commits himself exclusively to one specific preconceived theory . . . for then he becomes doctrinaire and can no longer "see around" either his pet theory or any other. He becomes insensitive, or even defensive, toward the kinds of questions that cast doubt on his theory; he is preoccupied with testing, modifying and seeing everything from this one angle (Glaser & Strauss, 1967, p.46).

This passage assumes that "committing oneself exclusively" to one "angle" implies rejection of others and refusal of contradictory evidence. This is perhaps what we might expect of such "exclusive commitment." But if it is a case of "examining" rather than "exclusively committing oneself to" a particular theory, then we might expect a more open approach, both to the use of evidence and the potential purchase of other theories.

Criticisms of the Idiographic Approach

As grounded theory is quintessentially a comparative method, it is not surprising that Glaser and Strauss, at least in their original text, are critical of the single substantive study. For them, theory is born through comparison of (empirical) similarities and differences across substantive areas. Therefore the single study has at best a very limited role, since without taking into account "all the contingencies and qualifications" of other substantive areas it is "too sparsely developed" (Glaser & Strauss, 1967, pp.81–2). In discussing the use by Geertz of two comparison groups, for example, they suggest that the resulting theoretical framework is "really quite thin" (Glaser & Strauss, 1967, p.147). In the interests of generating comparison, they are more inclined to use 40 groups than four.

The reason that Glaser and Strauss need to study so many substantive areas is that comparison in grounded theory is empirically generated rather than theoretically driven. Whatever the empirical riches that can be mined from the study of a particular substantive area (cf. Stake, 1994, pp.236–247), on its own it cannot provide useful comparisons. However, theoretically driven case studies may have more to offer. The case can be, and usually is, seen as a "case of" something else (Walton, 1992, p.121). Walton takes this meaning, with its "hint of generality," as self-evident, but the term "case" can also refer to something which is "encased" (which is how Stake uses it). By "encased" I mean that the phenomenon is studied in its own right and not as an example of anything else.

Two Types of Case Study

- *In a "case of" study the case is selected and studied as an example of some wider population or phenomenon*
- *In an "encased" study a case is selected without reference to any wide population or phenomenon, but studied in its own terms*

Given this ambiguity, the term "case study" is virtually meaningless, since it can be interpreted in such different and, indeed, contradictory ways. It is notable that Glaser and Strauss (and Strauss and Corbin) studiously avoid any reference to "cases" or "case studies"—perhaps because, from their perspective, neither an "encased" study nor a "case of" study can generate empirical comparisons.

"Case of" Studies

What a "case of" study can do, though, is generate theoretical comparisons, for in such studies, "cases come wrapped in theories" (Walton, 1992, p.122). These theories typically refer (at least implicitly) to a wider world which the case in some way represents:

> Case studies get at the causal texture of social life, but drift without anchor unless they are incorporated into some typology of general processes, made causally explicit within the case, and ultimately referred back to the universe which the case represents, at least hypothetically (Walton, 1992, p.124).

According to Walton, generalization from a case study can be done in at least two ways. The first is substantive and involves using a case to substantiate or expand on interpretations in similar cases (such as other hospital studies, for example). Here generalization obviously depends on the accumulation of relevant cases over time. The second is analytical and involves a strategic argument justifying the case as critical from some theoretical angle—for example, because it represents a "deviant" case. For this purpose, we may need to compare the particular case with a theoretical model of the "typical" case, against which differences can then be judged (Ragin, 1987, p.39). Sayer (1992, p.249) argues that "rare conjunctures" can often "lay bare structures and mechanisms which are normally hidden." Both the substantive and analytic approaches add weight to a "case of" study by locating it within a wider (theoretically defined) context or investing it with strategic theoretical import—but neither of these options are countenanced in grounded theory.

Walton adds a third way of using the "case of" study, even more relevant given the emphasis in grounded theory on generation rather than verification of theory. Though Walton does not name this approach, we might call it "innovative." The idea is that these studies generate new ways of thinking about the world from which the case is drawn. We see things in a new way, rather as Goffman's work transformed the mental hospital into a "total institution." Thus some "case of" studies "provide models capable of instructive transferability to other settings" (Walton, 1992, p.126). In these studies, "the universe is inferred from the case":

> The older universe, itself an expression of theory, is disaggregated and some of its elements recombined with newly perceived phenomena in a universe reconstructed as the field of new explanations (Walton, 1992, p.126).

The new explanations succeed "because they supplant previous ones—they explain the old facts and more." Walton describes these as "small revolts" in the research process; but though "small," by reformulating received ideas regarding causal processes, they can provoke "a rash of studies that replicate and improvise on one case." So "case of" studies can result in theoretical innovation, rather like the paradigmatic revolutions of Kuhn, though "less abrupt and discontinuous" in character (Walton, 1992, pp.126–7). These points may prompt us to pause before writing off the "case of" study as a theoretical dead-end.

Generalizing from "case of" studies

- *Substantive generalizations can be made through comparison with other (previously studied) cases*
- *Analytic generalizations can be made through comparison with a theoretical model of a typical case*
- *Innovative generalizations can be made through the reformulation of received ideas*

"Encased" Studies

Let us turn to the "encased" study. This refers to the study of a particular substantive area, where the aim is not to generalize from the particular but to understand it as a unique and complex phenomenon. Whereas the "case of" study is nomothetic in its ambitions, the "encased" study is idiographic. But this is not to deny it theoretical import, unless we (mistakenly I think) equate theory with generalizations. The "encased" study may go a long way toward explaining the particular phenomenon and understanding its implications. For example, there is only one global economic system (unless we subscribe to the view that multiple interpretations create multiple realities, in which case there are over six billion!) but an encased study of it may produce explanations of some import. This is not because we want to generalize about global economies, but rather because understanding this particular global economy may contribute to wider theories of how society works (and where we go from here).

Thus it is the wider picture that gives the encased study its theoretical import, rather like the addition of another jigsaw piece may help us to complete a puzzle. As I suggested earlier, we may work on large or small pieces or even more than one piece at a time. This analogy also suggests that, from the point of view of generating theory, some care needs to be exercised over which piece(s) of the puzzle to work on. If researchers select trivial or isolated pieces, while assuming, ignoring, or discounting the wider picture, the encased case can hardly have much theoretical import. Here again, however, we need not disown a potentially powerful procedure simply because of bad practice.

Incidentally, it is doubtful whether one could in any case grasp the theoretical import of one piece of the puzzle through generalization. While generalizations may help us to understand "where things go," they do not capture the complex connections between pieces of a puzzle. For example, suppose I am working on a puzzle in which a dove symbolising peace has just been released into a blue sky. Generalizations ("this bit is blue") may help me sort out the pieces but they do not convey their significance in the overall configuration. The piece with the white dove (this is an easy puzzle) transforms the picture, which is not an aggregation of separate parts but a complex whole. And the whole picture is also implicit—or "implicated (Bohm, 1983, p.149)—in each separate part: we cannot separate the significance of the piece with the dove on it from that of the picture as a whole.

We are used to thinking that wholes contain parts, but are less familiar with the idea that parts also contain wholes (Kosko, 1994, p.58). It is usual to regard wholes either as aggregates of separate parts or as structures into which the separate parts fit. In these conceptions, each part can be conceived separately from the whole. However, to complete a puzzle, the parts cannot be conceived separately, but only as partial expressions of the whole picture. We could say that the whole picture is inherent or "enfolded" (Bohm, 1983, p.149) in the encased study, rather as a cash economy—and indeed a whole social order—is implicated in my trip to the shops. If this is so, then the theoretical import of the encased study may lie in how the whole can be illuminated through understanding the part and not just (more conventionally) in how the part can be illuminated through understanding the whole.

The rules that govern this kind of thinking may be neither inductive (going from the particular to the general) nor deductive (going from the general to the particular). The connections between a part of a puzzle and the whole may based on various modes of thinking. Returning to the example of the dove, we can recognize metaphorical reasoning in its role as symbolising peace. This is not derived from empirical generalization (we do not find a constant conjunction of doves and peace). Given that a dove typically symbolizes peace, we can identify how it contributes to the effect of the overall picture. This is not through deductive reasoning, however ("if dove, then peace"), but through an understanding of how the symbolism produces the desired effect or what Sayer (1992, p.107) describes as a form of "retroductive" reasoning about causal mechanisms.

CONCLUSION

Let us take stock.

We have questioned the distinction between substantive and formal theory, in terms of levels of generality. We have distinguished intensive and extensive forms of theorizing in terms of their orientation—to explain particular phenomena (in their unique complexity) or to produce generalizations. It appears that grounded theory conforms to neither of these forms of theorizing, but offers a hybrid approach. This combines elements of intensive and extensive theorizing, but may incur some costs in so doing. In grounded theory the emphasis on comparison across a range of "areas of inquiry" (not cases) may preclude effective study of any particular phenomenon. The emphasis on integration through "explicating a story line" and "theoretical sorting" is intended to support a complex, multivariate account—but one that reduces rigor by relaxing the canons of comparative inquiry. It is not clear, therefore, whether grounded theory offers us the best or worst of both worlds. If the best, it offers a way of producing generalizations through comparison while retaining a holistic sense of causal complexity. If the worst, we lose both the depth of intensive theorizing and the rigor of comparative inquiry.

However, the opposition of intensive and extensive forms of theorizing may be less clear-cut than assumed by their advocates and critics. Here I followed Wallerstein in distinguishing between nomothetic theorizing, which seeks universal generalizations, and idiographic theorizing, which focuses on particular events. I suggested that nomothetic theorizing can only be done within a given (if unspecified) space-time context. I also suggested that idiographic theorizing relies on generalizations. If so, the divide between these modes of theorizing is not as profound as sometimes imagined. This perhaps provides further encouragement to pursue the kind of mixed methodology adopted in grounded theory. On the other hand, it also suggests that we need to reconsider the criticisms that are directed from a grounded theory standpoint against other forms of theorizing.

The rationale for grounded theory lies partly in a (more or less veiled) critique of the generative power of other modes of theorizing. Glaser and Strauss are especially critical of speculative/deductive theorizing, preferring to derive generalizations from empirical observation. Their argument proceeds, though, through a series of conflations—of "ungrounded" premises and logical deductions; of deductive reasoning and "forcing" data; of speculative and undisciplined theorizing. The case for basing nomothetic theorizing entirely on empirical inquiry seems rather weak. The case against idiographic studies also underestimates the contribution these can make to theorizing. Here I distinguished between "case of" studies, where a case is studied as an example of some generality, and "encased" studies, where a case is studied in its own right and not as an exemplar. Although neither "case of" nor "encased" studies alone provide a basis for generalizations, this is not the only mode through which theory can be generated. The "case of" studies can generate theory through comparison with other studies or theoretical models or through generating innovative perspectives on how to theorize the generalities that the case represents. The "encased" study can generate theory, not by generalizing from but through illuminating connections within a particular case—how pieces fit within an overall puzzle.

On the other hand, grounded theory offers the prospect of a hybrid methodology that transcends the divide between intensive and extensive forms of theorizing, promising to combine the strengths of both. On the other hand, our reappraisal of idiographic and nomothetic forms of theorizing suggests that the particular methodological mix which grounded theory offers could be strengthened. There seem strong arguments for accepting elements of both nomothetic theorizing and of idiographic theorizing (such as theoretical speculation, logical deduction, and case studies) as part of the theorizing strategies that can be utilized within a grounded theory approach.

CHAPTER 10

Making Theory Fit

As grounded theory originated in a vigorous and uncompromising critique of verificational studies, it may seem rather paradoxical to devote a chapter of this book to verification in grounded theory. However, some enemies just refuse to accept defeat; and so it is with verification. We have already seen at various points how Glaser and Strauss find it difficult to follow their own remit, and—despite their denunciations of verification—continue to make claims that imply at least some form of verification for grounded theory. If anything, such claims for grounded theory have since become more insistent with the advent of software supporting the comprehensive coding of data, with attendant calls for greater rigor in qualitative research. Given this claim to greater rigor, we need to reconsider the claim that theory can be grounded through the very process through which it is generated.

THE AIMS OF THEORIZING

A useful starting point in considering whether theory fits the objects of our inquiry is to consider what theory has to be fit for.

In grounded theory Glaser and Strauss (1967, p.3) describe the general aims of theorizing along the following lines:

- To enable prediction and explanation of behavior
- To contribute to theoretical advance in sociology
- To produce practical applications—prediction and explanation should be able to give the practitioner understanding and some control over situations
- To provide a perspective on behavior—a stance to be taken toward data
- To guide and provide a style for research on particular areas of behavior

According to Glaser and Strauss, these tasks require theory that must meet two criteria: first, that it "can be verified in present or future research;" and, second, that it "must also be readily understandable" to "significant laymen" as well as sociologists. This in turn requires theory "which must fit the situation being researched, and work when put into use":

> By "fit" we mean that the categories must be readily (not forcibly) applicable to and indicated by the data under study; by "work" we mean that they must be meaningfully relevant to and be able to explain the behavior under study (Glaser & Strauss, 1967, p.3).

Later Glaser and Strauss add some further criteria. Theory must be "sufficiently general to be applicable to a multitude of diverse daily situations within the substantive area" and it must "allow the user partial control over the structure and process of daily situations as they change through time" (Glaser & Strauss, 1967, p.237).

As theory is applied to an "ever-changing, everyday reality" (Glaser & Strauss, 1967, p.239) it is important that it "fit" that reality—rather than reflecting the researcher's own values as embodied in a formal theory which is applied to the situation (through logical deduction) without regard to its relevance. Applying theory through deduction results in theories "divorced from the everyday realities of substantive areas" leading to "neglect, distortion and forcing." This woeful catalog of sins can be avoided if theory fits because it is induced from observation and therefore "faithful to everyday realities" (Glaser & Strauss, 1967, pp.238–9).

But how do we know if a theory fits? Glaser and Strauss seem ambivalent in their response to this question. On the one hand, they seem to assume that a theory fits if it has been induced from observation. Hence their claim that "the adequacy of a theory for sociology today cannot be divorced from the process by which it is generated," for a theory "is likely to be a better theory to the degree that it has been inductively developed from social research" (Glaser & Strauss, 1967, p.5). In general, they treat observation as an atheoretical tool, relatively untrammelled by the selective biases and preconceptions of the observer. Observation allied to comparative procedures can therefore provide a solid basis for inference. Here theorizing seems driven by whatever is in the data. On the other hand, they also admit that the generalizations produced through grounded theory may still require verification.

This is because the canons of comparative inquiry are deliberately relaxed in the interests of generating theory—making subsequent verification essential.

On balance, however, Glaser and Strauss seem content to rest on their laurels—that is, to claim that a theory will fit sufficiently if it has been grounded in comparative inquiry. This inclination is reinforced to the extent that they incorporate verification into grounded theory—usually as junior partner but on occasion even equating the two:

> . . . it is the purposeful grounding or verification process that makes this mode of theory building different from many other modes of theory building. . . (Strauss & Corbin, 1990, p.112).

The "verify as you go" approach tends to win out against the "verify later" perspective—despite the frequent use of methodological procedures which are justified precisely because theory is being generated rather than verified.

In any case, time is too short, they argue, to entertain doubts at all seriously, so that theory typically "never gets to the stage of rigorous demonstration" (Glaser & Strauss, 1967, p.235). Events (and theorists) move on—though it then becomes less clear how generating theory can contribute to "theoretical advance" with its implied accumulation of knowledge based on earlier theories. The emphasis shifts to discovering and exploring change rather than accumulating a stock of knowledge that can withstand the tests of time. Quite where this leaves the fit of formal theory (that is, theory at a higher level of generality) is less than clear.

In their methodological overview of grounded theory, Strauss and Corbin (1995) suggest that theories are always provisional (subject to continual elaboration and qualification) and limited in time. As conditions change, so does the relation of theory to contemporary reality. In grounded theory adaptation to new conditions is inherent in the method of starting afresh with each new inquiry. It is also facilitated by the production of a theory that is conceptually "dense" and therefore sensitive to the range of (changing) conditions under which it can be applied. Predictability from this standpoint means only that "if elsewhere approximately similar conditions obtain, then approximately similar consequences should occur" (Strauss & Corbin, 1995, p.278).

From this perspective, the test of a theory lies less in its general truth than in its practical adequacy in particular circumstances: "the application is thus, in one sense, the theory's further test and validation" (Glaser & Strauss, 1967, p.244). Hence the stress in grounded theory on theorizing in a way that is comprehensible to respondents and practitioners (and not just researchers). Theory allows prediction and offers control by illuminating the connections between conditions, interactions, and consequences—but only in particular contexts.

It is less clear, though, how these contexts are specified. The function of theoretical sampling is to generate theory through comparison and not to delineate in any detail the contexts within which interaction occurs. It is sufficient for theoretical sampling that situations produce useful comparisons for the categories under investigation—there is no attempt to study each case as a whole. Indeed, the spatial and temporal context is to some extent marginalised by a mode of theorizing that eliminates cases from the research vocabulary. The task of determining relevant contexts is relegated to those who want to apply theory—they will "make the necessary corrections, adjustments, invalidations and inapplications when thinking about or using the theory" (Glaser & Strauss, 1967, p.232). Thus a grounded theory is continually adjusted in the process of its application. Yet it still retains sufficient generality to apply across a range of contexts, and a sufficiently holistic view to convey "the total picture" (Glaser & Strauss, 1967, p.243).

PROBLEMS WITH PREDICTION

Meeting these tasks is, as the authors put it, a "large order" (Glaser & Strauss, 1967, p.3), particularly as theory has also to be flexible, given the fluidity of social change. How are these ambitions to be realized?

If the application of grounded theory (that is, prediction and control) requires "the total picture," then perhaps it cannot be. For as Sayer (1992) argues, it is difficult if not impossible to emulate in social science the kind of closed systems that make prediction possible in some (but by no means all) of the natural sciences. Prediction requires not only an explanation (along the lines of "if A, then B") but also a specification of the conditions under which this holds true. This is possible in closed systems where "objects and their relations are stable," so that the state of the system can be fully described; but "the possibilities for accurate and reliable explanatory predictions for open systems are remote" (Sayer, 1992, p.131). It is simply not possible to specify in advance the requisite conditions. Explanations that tell us "what makes things happen" cannot readily be translated into predictions that give us "grounds for expecting something to happen" (Sayer, 1992, p.134). This is not just a problem for social sciences, as the natural sciences "despite their mythology, share many of the same problems we have" (Fischer, 1994, p.82). But one particular reason for uncertainty in social sciences lies in our capacity to initiate change. We are not content to predict events—we make them happen. The future is shaped by our current interactions.

Sayer suggests that this makes prediction on the basis of invariant patterns of events (or "regularities") rather awkward. But it does not render prediction entirely impossible, if we can at least understand the possibilities inherent in the exercise of particular powers and processes. Thus we can predict with confidence that a new government will take office after an election—though not which party will be in power. We can make this prediction because the election is a process that culminates (in the future) in the formation of a new government. Even this prediction depends on some contextual assumptions—most notably, that the military will not intervene to forestall a result they find uncongenial! However, this "prediction" may at least alert us to what must be done (that is, what possibilities must be realized) in order to produce a certain result.

The future may not be predictable in a precise way, but neither is it arbitrary. Between these polar positions, we may recognize a range of possibilities. We tend to think that a thing is either possible or it is not. But, strange to say, some things are more possible than others. For example, it is still easier to shop on foot in urban Scotland than in the urban centers of North America. That is partly because the Scottish urban areas are older and more highly concentrated, and (as yet) fewer shopping malls and supermarkets are located in peripheral areas accessible only by car. In Scotland, people are still expected to walk and the means (pavements, or sidewalks) to do so are provided. But as the supermarkets and malls develop, the local and specialist shops may dwindle in number, the traffic volume increases, the roads widen, the pavements disappear—and the possibilities for shopping by foot diminish in consequence.

The possibilities of prediction also vary, depending on the extent of our knowledge of powers and processes and the conditions under which they may operate. Where structures (such as electoral systems) are fairly entrenched, with established rules that insist on certain interactions (polls every four years, for example) and prohibit others (such as ballot rigging), possibilities are constrained (and prediction accordingly easier). Where structures are less well established, we have to take more account of the possibilities of their transformation. While the interplay between structure and agency may be unpredictable (because our own actions determine what is reproduced and what is transformed), it is not entirely arbitrary either. As we have seen, agency may be shaped by situational logics or the distribution of interests and the power to realize them. In other words, social situations are rarely "closed" (and so entirely predictable in principle if not in practice); but they can be considered more or less "open."

Take, for example, the phenomenon of "credentialism"—the inflation of educational qualifications. Suppose we expand educational opportunities rapidly, outpacing any increase in the number of desirable jobs through corresponding expansion or change in labor market structures. Insofar as educational qualifications are used to assign people to places in the queue for desirable jobs, we can expect these jobs to require higher credentials over time, regardless of any changes in their skill content. The structural change (education expands while employment is static) brings a new situational logic. For employers, raising qualification levels provides a way of screening potential candidates as competition increases. For students, obtaining higher qualifications allows them compete more effectively for the more desirable jobs. This in turn creates new pressures for educational expansion. One result may be graduate unemployment, as too many people compete for too few places. Another may be underemployment, as people take jobs for which they are overqualified. Who becomes unemployed or underemployed may reflect the existing opportunity structure, which eases the path of some while creating obstacles for others. In the longer term, though, the labor market may adjust to a more qualified intake, as employers also have to compete for the best applicants, which they may do by making the jobs they offer more attractive; to remain competitive, they may also want to exploit more fully the human resources at their disposal.

This theory "works" because it deals with a situation that is not entirely open. Some things are taken as given—including the expansion in education, the lack of desirable jobs, and (to a lesser extent) the competition for them. At a more basic level, we look to employment to generate the resources we need to live in the absence of acceptable or feasible alternatives. Other things—such as competitive bias—may vary depending on the extent of existing inequalities or policies pursued (such as affirmative action programs) that may increase or reduce it. The degree to which a situation is "open" depends in part on how amenable these basic constraints are to manipulation and change—in other words, on the time-scale for potential change in structural conditions, relative to the time-scale required for interaction. We cannot hope to make significant changes in the ratio of desirable jobs to educational opportunities in the short run. And in the long run, of course, we are all dead.

DIFFERENTIATING POSSIBILITIES

If, rather than prediction, practical control depends on an assessment of possibilities, then how can we differentiate them?

One approach involves qualitative comparison of the possible conditions across a number of cases (Ragin, 1987). This involves identifying different configurations of conditions associated with each outcome and then comparing each configuration across the range of cases. These comparisons can be formalised using the (Boolean) algebra of logic and sets to produce a summary account of those combinations that are most closely associated with outcomes (Ragin, 1995, p.182). In this way we can identify among a range of possible causal combinations those that seem most closely implicated in outcomes. As causes are not analyzed in isolation, this form of analysis is intended to extend the methods of agreement and difference to include multiple and conjunctural forms of causation. However, it still does so by analyzing the associations between events—it relies on identifying patterns among causal processes rather than rendering intelligible the application of causal powers.

I have suggested that time-scale may be another consideration. I do not mean that the more enduring a constraint has proved, the more possible its reproduction and the less likely its transformation. For one thing, this is too retrospective (the past may be a poor guide to the future). We may have had 100 years of peace (or war) but that in itself offers no guide to the prospects of change. On the eve of World War I in Britain, it would have been a poor prediction that peace would continue because the country had not been involved in a continental war since the middle of the previous century. And wars (no matter how long) can also end abruptly and unexpectedly in sudden victory or inglorious defeat. The past is only a guide to the future if we can comprehend the reasons why the future may be (un-)like the past. This depends not so much on the projection of existing trends as on understanding the processes that produce them.

The more basic these processes are, the more confidence we may have in our understanding of them. We know, for example, that people need to eat; therefore they need an income; therefore they need jobs; therefore they need qualifications; therefore they need an education. This is a long line of reasoning, that only makes sense within a given context (societies where alternative sources of income are limited and jobs are allocated on the basis of educational qualifications). No doubt it is subject to qualifications—there may be other sources of income (such as social security or self employment), for example, and people may value education for its own sake and not just as a means to an end. But we can be sure that people need to eat—and communicate. It is this substratum of "basic needs" (however defined) that can add weight to explanation and prediction. As Lakoff (1987, p.299) argues, we can be more confident about our knowledge at a basic-level. This also implies greater confidence about our knowledge of the structures and interactions that impinge directly on basic-level experience—food, shelter, sex, entertainment, communication, reproduction and the like—than about more esoteric forms of knowledge (which may rely on metaphorical and metonymic projections from the basic-level). The "diffusion" of technology or the economic "cycle" are examples of such projections, and here our knowledge is less confident because we lack a direct understanding of the processes to which these concepts refer.

Another way of differentiating among possibilities is in terms of an evolutionary perspective (Kumar, 1994). We do not live and act in a vacuum, but as a result of a long series of evolutionary adaptations which open up some possibilities while forestalling or foreclosing others. Educational systems are themselves the result of an evolutionary process, and as such forestall some alternatives (learning through work) while foreclosing others (learning through oral tradition). From an evolutionary perspective, we can ask about the fitness of any adaptation given the changes which may have occurred in the environment.

First we have to discount the misinterpretation of Darwinian theory as implying "survival of the fittest." This took little account of the vital role of chance in the evolutionary process and failed to recognize that outcomes can be highly sensitive to even slight changes in initial conditions. Feedback processes can consolidate and amplify the effects of an initial configuration, just as student pressures for qualifications can amplify credentialism. Organisms can become locked in to a particular process (like an arms race) which may ultimately threaten adaptation to a changing environment. Evolutionary paths therefore contain pitfalls and cul-de-sacs from which organisms may suddenly have to retreat in order to reach conditions allowing survival or further progress. Therefore evolutionary change tends to involve long periods of gradual consolidation followed by periods of crisis and rapid change.

From this perspective, we can differentiate possibilities by analyzing evolutionary paths both to assess the extent of fitness with a changing environment, and to identify the degree to which we are locked in to particular processes. The first of these focuses on harmony or conflict between agency and structure among agents or within structures themselves. What are the benefits and drawbacks for agents of the ever lengthening educational process and how significant are they? How does competition affect the learning process, if this requires cooperative endeavor? Does educational expansion impose unnecessary costs on industry, undermining rather than enhancing competitiveness (at least in the short term)? The extent to which different agencies, structures, and interests complement, or contradict, each other may give some clues as to the prospects of change. However, it should not be assumed that conflict and contradiction automatically engender change, while symbiotic relationships spell stability. As with the arms race, it is equally possible to become locked into conflicting relationships.

This issue (of lock-ins) raises the question of what feedback mechanisms exist; and whether and how these reinforce or reduce a particular process once initiated. Reverting to my earlier example, the location of supermarkets and malls in peripheral areas accessible only by car creates feedback mechanisms (once you have a car you may as well use it, local shops are depleted, etc.) that reinforce expansion—creating a new problem in the form of traffic congestion. Educational expansion also creates reinforcing pressures through qualification inflation; but there may also be mitigating pressures, particularly in the costs imposed by extra education on time and resources. These examples refer to a calculation of costs and benefits, but remember that lock-in may result from their benefits at your cost (or vice versa). Ultimately no one may benefit—the depletion of fishing grounds being a notable example of an industry locked in to an ultimately self-defeating process, in the absence of a commercial structure capable of curtailing overfishing. As in this case, it may take an ensuing crisis to produce a response which confronts (or perhaps succumbs) to the basic constraints under which the industry operates.

I have presented lock-ins with rather negative examples, reflecting perhaps the everyday connotations of being locked into a particular course of action. But we can be locked into virtuous as well as vicious circles (or spirals). Welfare states are sometimes cited as incentives to virtue, since they create arrangements where one can invest in one's own welfare (by paying taxes for education and health services, for example) while enhancing the common good. In some respects, even "qualification inflation" might be regarded as a virtuous spiral, as educational expansion has the largely unintended consequence of extending the experience of higher forms of education to a larger proportion of the population. If education brings various intrinsic and extrinsic benefits—such as a reduction in gender inequalities, improvements in child health, and so on—then these consequences may perhaps more than offset the short-term negative effects of unemployment and underemployment.

The converse of lock-ins are those situations where a new configuration emerges, perhaps because some interruption in the feedback process or change in conditions permits escape from the constraints prevailing hitherto. We can call these situations "break-outs" (though again we should be wary of the connotations, this time positive). Perhaps the most dramatic breakouts are to be found in revolutions, economic as well as political. The postwar Keynsian revolution represented a break-out from the stagnant conditions of the inter-war years (at least in Britain) until undermined by inflationary pressures and global competition. The formation of the European Union represents (at least for some) an attempt to break out from the wars that have so marred the history of Europe. The most dramatic break-outs may not be the most profound. One of the most significant break-outs of this century has surely been that of women, who, thanks to technological innovation, are no longer locked into a cycle of childbirth and marriage.

The credentialism theory allows us to explain why rapid educational expansion in African countries resulted in high levels of graduate unemployment and underemployment—though it may have had offsetting social and political benefits. We can also predict similar consequences of any rapid expansion of educational opportunities relative to the current stock of desirable jobs. However, the point of the understanding credentialism is often (as in this case) not to predict these results, but rather to consider how if possible to avoid them. This is comforting because it allows control to be exercised on the basis of explanation rather than prediction. Given some understanding of powers and processes, we can recognize the possibilities inherent in a situation and act accordingly.

From the standpoint of control, then, we are less concerned with predicting events than with understanding what produces them. As we have seen, such understanding may be based on an analysis of particular configurations (idiographic theorizing) as well as on generalizations (nomothetic theorizing). Of course, understanding may not confer control (with its mechanical overtones) so much as guidance which can inform choices. Ironically, we seek such guidance less where events are predictable (usually because they are routine) than where options are available and a choice is required. In other words, it is precisely where events are unpredictable (but explicable) that we most rely on theorizing to inform our choices (Sayer, 1992, p.136).

Some Ways of Assessing Possibilities for Interaction

- *Through comparative analysis across cases*
- *Through assessing the time-scales involved in reducing or eliminating constraints*
- *Through assessing degrees of fitness to an environment*
- *Through identifying lock-ins and break-outs*

Though Glaser and Strauss emphasize the role of grounded theory in providing guideliness for practical action, they offer little advice as to how theory can be used to inform choices. I have suggested that we need to think less in terms of prediction and more in terms of identifying and weighing possibilities. A number of considerations may weigh with us in assessing possibilities for interaction. We can compare possibilities through comparative analysis across a range of cases. We can compare the time-scale involved in changing constraints with the time-scale for interaction. This requires understanding of (that is, theorizing about) the processes involved. I have suggested that we may be more confident about our understanding, the closer these processes impinge on our basic experience and knowledge. We may consider the extent to which a process is adapted to its environment and whether these complement or contradict each other. We can also look for lock-ins and break-outs: the extent to which agents are locked in to existing processes or can break out of them. In these (and no doubt other) ways, we can differentiate among the possibilities for interaction without predicting particular events—or assuming that we must conform to prevailing patterns.

UTILITY AND VALIDITY

If the test of practical adequacy does not lie in prediction (except, perhaps, in special circumstances which approximate to closed systems) then how can we verify our theories? We might argue, along with Glaser and Strauss, that they are tested sufficiently if they prove their value in practical applications—that is, if they allow us to exercise some effective control over events. In other words, theory is practically adequate if it proves a good guide to action—if it is useful.

Glaser and Strauss seem to flirt with this "instrumentalist" position, though (since they want theories which fit as well as work) they do not endorse it altogether. One problem with an instrumentalist view is that it abandons any test of the truth of a theory, in favor of assessing its utility—a position which has been described by some as "untenable to any intellectually honest scholar" (Sylvan & Glassner, 1985, p.105). For a theory may be useful without being true—just as eugenic theories, though false, were (and are) "useful" in justifying racism.

One obvious issue here is: useful to whom? The utility of a theory may vary according to the interests of those affected by its application. We do not (and cannot) expect any agreement on utility—we expect it to vary according to person and circumstance. The truth of a theory, in contrast, is something we can and do expect to agree about, independently of person and circumstance. We do not accept a theory as true just because it is (personally or momentarily) convenient. Nor do we hold something to be true just because it is conventional or widely believed. Truth is not a matter of convenience, convention or consensus, but of constraint. We are obliged to accede to the truth of a theory by the force of evidence. This does not mean that the truth of a theory is absolute (for we are not omnipotent)—but we can reject it only by producing another (and better) account of the evidence it explains.

If this is so, then we are also obliged to consider the conditions under which theory can claim to be "grounded." Although the strongest claim made by Glaser and Strauss is that theory can be verified, either in the process of its production, or by further testing, this perhaps poses too stringent a condition, if it were taken to mean establishing truth beyond any possible doubt. The problem here is that there may always be more than one theory consistent with the available evidence. Indeed, Eddison once claimed to have constructed three thousand different theories about electric light, each one "reasonable and apparently likely to be true"—and to have disproved all but two by experiment (Gleick, 1994, p.319). As experiments are not so readily available (if at all) to social science, such a fecundity of theories would be even more of an embarrassment. At any rate, we cannot hope to prove theories true—or even, for that matter, to prove them false (despite Popper's claims for falsification). The best we ca hope for is not to verify or falsify theory but to produce evidence strong enough to allow us to confirm or reject alternative accounts.

Verify: Establish the truth or correctness of, by examination or demonstration.
Confirm: Establish more firmly, corroborate.

Therefore it seems more appropriate to speak (as Strauss and Corbin sometimes do) of validating than of verifying theory. To validate is to confirm, and confirmation is as close as we can get to establishing the truth of a theory. In an earlier text, I defined a valid account (or theory) as one "which can be defended as sound because it is well-grounded conceptually and empirically" (Dey, 1993, p.228). If (but only if) we take the claim of grounded theory to be grounded at face value (that is, as not requiring any subsequent validation), then we have to ask on what grounds it can claim to produce a valid account.

THE DISCIPLINED IMAGINATION

One of the great strengths of grounded theory undoubtedly lies in its determination to discipline the discovery of theory by reference to research. Given the profligacy of social theorists and the paucity of the empirical research, this seems a laudable aim. I suggested earlier that speculation (and logical deduction) also has a place in generating theory, but only if disciplined by certain constraints. Research offers an anchor for speculations that might otherwise simply become flights of fancy. However, the relationship between research and theorizing in grounded theory is perhaps less stringent than might first appear.

In the first place, in grounded theory the main discipline on theorizing is that imposed by the data produced in the current project. Little explicit account need be taken of the evidence produced by other research. The theory generated need only be valid for the situations included in the study, for there is no suggestion of cross-checking the ideas produced in a grounded theory study against data produced in other research. If we were to insist that the emerging theory should be consistent with evidence already available from previous studies, we might have stronger grounds for claiming validity for the theory generated through this research.

I suggested that a valid theory has to be well-grounded conceptually as well as empirically. If an emergent theory is inconsistent with other theories, we may reasonably doubt its validity—most especially where we have strong grounds for supporting the other theories. In grounded theory we are under no obligation to consider the consistency of our theory with other theories. Quite the contrary: we are advised to take no account of prior theory, most especially where it bears directly upon the field under investigation. This procedure is intended to free us from preconceptions. Whatever its merits in that regard, it does free us from any need to consider the (in)consistency of the emerging theory with other theories. But we might have better grounds for accepting an emerging theory as valid, if it can be shown to be consistent with other theories.

GROUNDING THEORY CONCEPTUALLY

A theory is not just a haphazard collection of concepts—the ideas in a theory are related in a systematic way. To be well-grounded conceptually, the relationships between concepts used in a theory must be set out systematically. The connections between concepts need not always be logical (as Lakoff demonstrated) but they should at least be clear. In grounded theory we have encountered various confusions, arising from the problems of applying the methods of agreement and difference, for example; or the difficulties attendant on combining intensive and extensive methods of study. Many of these problems stem from attempting to analyze causality in terms of constant conjunction within a framework that purports to offer a more complex and holistic account. As a result, the grounds upon which connections can be established between concepts are far from clear.

Theory: Supposition or system of ideas explaining something, especially one based on general principles independent of the facts, phenomena, and so on to be explained; speculative (especially fanciful) view; the sphere of abstract knowledge or speculative thought; exposition of principles of a science; collection of propositions to illustrate principles of a subject.

If evidence can always support alternative accounts, it behooves the researcher to consider the strengths and weaknesses of rival interpretations. But in grounded theory there seems to be no such thing as an alternative interpretation. There is only room for one account. This preoccupation with the production of a systematic but singular account is most evident in the (if need be, arbitrary) selection of core categories as the central theme of the research. Even where research is conducted by a team, debates contribute only to the development of a "shared" analysis (Glaser & Strauss, 1967, p.226). While the lone researcher may struggle through a series of partial analyses, this merely clears the ground for the emergence of the "final analysis" (Glaser & Strauss, 1967, p.225). The possibility of alternative interpretations is recognized in principle, but relegated to the bottom drawer in practice. Whatever alternative accounts may have been considered, they are destined never to see the light of day—at any rate, before the next publication.

This places the research community at a severe disadvantage when it comes to assessing the validity of the resulting report, for it is presented with a "fait accompli." To establish credibility, the researcher is encouraged to summarize their main theoretical framework, describe their data vividly (through judicious use of quotes, etc.), and explain their procedures—lest their months or years of "hard study" be dismissed as "somewhat impressionistic" (Glaser & Strauss, 1967, pp.228–9). But this is hardly enough. At the very least, there needs to be some attempt to identify ambiguities, exceptions, and errors, to assess the strength and quality of evidence, and to explain the origins and evolution of the concepts that are used—so that the research community can make its own assessment of the analysis and, if need be, draw its own conclusions from the research. In other words, there should be evidence enough to allow the research community to entertain alternative interpretations. It would be better still if the researchers themselves attempted to identify and assess alternative accounts.

Some Ways of Grounding Theory Conceptually

- *Consider its consistency with other theories*
- *Clarify the connections between concepts and the grounds for inference*
- *Identify errors, ambiguities, and exceptions in the analysis*
- *Assess alternative explanations consistent with the data*
- *Provide an audit of the emergence of theoretical ideas*

This is an area where computer software has transformed the possibilities for qualitative research. The opportunities for greater rigor and transparency in the management and analysis of data provide a basis for grounding theory more effectively in the available evidence. And the available evidence need no longer be confined to some choice quotations, some judicious illustrations, and the occasional case profile. Improved storage and transfer facilities may allow access to the original files on which the analysis is based.

Grounding Theory Empirically

Finally, we have the problem of grounding theory empirically. I have already suggested that the available evidence need not be confined to that produced through the current research study. But what of the way that data are produced and used in grounded theory?

The first issue here is with observation, since this at times tends to be presented as separate from theory, as though observation can be a neutral, preconceptual act. In grounded theory data are collected or gathered rather than produced. In discussing documentary sources, for example, Glaser and Strauss seem to place little emphasis on problems of achieving an adequate interpretation of the meaning of texts (for a striking contrast, see Forster, 1994). The researcher is recognized as playing an active role in generating theory, but this is only at the level of making hypotheses about relationships between concepts rather than in production of data or even, it sometimes seems, the concepts themselves. This may encourage a general lack of concern with observational bias on the part of the researcher, an indifference perhaps reinforced by the assumption that preconceptions can somehow be discarded at the outset of the research. But what we observe cannot be separated so easily from our preconceptions.

To take a topical example, I am writing this book in a country (Scotland!) currently sweltering in hot temperatures and high humidity. But it is even more oppressive in England, and today an English visitor stood on the doorstep and pronounced the weather wonderfully fresh and cool. Each observation of the climate was colored by experience. What we notice depends also on our background, inclinations, and values. To Glaser and Strauss, this presents no problem, for the researcher somehow rises above it all:

> . . . when different slices of data are submitted for comparative analysis, the result is not unbounding relativism. Instead, it is a proportioned view of the evidence, since, during comparison, biases of particular people and methods tend to reconcile themselves as the analyst discovers the underlying causes of variation (Glaser & Strauss, 1967, p.68).

In my example, the underlying cause of variation in climatic observations can be attributed to the different climatic experiences of observers from Scotland and England. But that is only one possible interpretation; and, moreover, one likely to appeal to geographers rather than physiologists. As physiologists, we might be inclined to attribute the differences in observation to physiological differences between the observers in their capacity to tolerate heat. Of course, this does not fit with the story as I presented it in terms of geography rather than physiology. But that is precisely my point: that my own observation (as honorary researcher) was not a neutral one, but one that already selected and emphasized certain aspects of the experience. Since observation is conceptually saturated, in a well-grounded analysis of data possible sources of bias in the production of evidence ought at least to be made explicit. Otherwise we are invited "to think with these hidden concepts but not about them" (Sayer, 1992, p.52).

As categories are central to analysis in grounded theory, we also need to think about categorization. Like observation, this is presented as basically straightforward and unproblematic. In grounded theory we categorize properties on the basis of similarities and differences—in conformity to what Lakoff calls the "folk theory" of classification. However, this is not how categorization actually works, if we accept the evidence on the role of prototypes, exemplars, schemas, and underlying cognitive models. If theory is to be well-grounded, we need to be aware of how we are using the tools of our trade. Moreover, if we accept the distinctions between substantive and formal categorization, and basic-level and more abstract categories, we also need to take account of different modes and levels of categorization.

Some Ways of Grounding Theory Empirically

- *Consider the (in-)consistency of the emergent theory with evidence from other research done in the field*
- *Make more explicit the conceptual assumptions that underpin observation*
- *Take account of different modes and levels of categorization*

There is nothing novel in this discussion of validating theory. Like the points raised earlier about conceptual grounding, these are simply the traditional criteria of validity, concerned with the logical structure of arguments, the fit between concepts and what is observed, the fit of the data with other evidence, and the consistency of the concepts used with those of other theories. These criteria are sometimes criticized as inappropriate for qualitative research, though it is striking how often they then simply reappear in a different guise (Mar & Rossman, 1989, pp.148–9).

These criteria are all concerned with internal validity—does the theory offer an adequate account of the evidence we have produced? We have said nothing so far about external validity—whether theory can reasonably be generalized beyond the data it purports to explain. This question hardly arises with idiographic theorizing, though in that context, we may question the use of generalizations to explain the particular. But it does arise with nomothetic theorizing, where we may question the use of generalizations derived from particular studies to explain the general.

Glaser and Strauss legitimate such generalizations by linking formal theory closely with the substantive studies, through which it is both generated and applied. Nevertheless, there seem to be problems with the use of grounded theory for generalizations. At least three such problems stand out. One is the preference to sample situations and processes rather than cases. This may make it difficult to locate the resulting theory in its local context, and take this into account when generalizing. A second is the use of theoretical sampling to select these situations and processes. This tends to leave open the question of how representative these may be. A third is the temptation to generalize without reference to the specific spatial and temporal context within which these generalizations may apply.

To these traditional criteria of validity, we can add more recent concerns, for example, with the role of the text (such as the conventions of plot and narrative) in structuring accounts and bolstering (possibly spurious) claims to authenticity (Altheide & Johnson 1994; Stones 1996, pp.167–191). This is not the place to engage with a positivist or postpositivist versus post-modernist or constructivist debate (a rather sterile one in my view) over the validation of accounts (Lincoln & Denzin, 1994). Some participants in this debate are prepared to dispense with validity altogether—and theory, too, for that matter. On the other hand, more reflexive forms of theorizing—that take account of the role and assumptions of the researcher, not only in producing data but also in producing interpretations and texts—may be seen as strengthening (rather than replacing) traditional criteria of validity.

If grounded theory is to live up to its name, then we might reasonably expect it to take account of these criteria. Nor would this be out of step with the spirit of grounded theory, which through its emphasis on explicit coding and systematic comparison has been responsible in part for the shift toward greater reflexivity and rigor in qualitative research.

Of course, the alternative logic of discovery may be called upon to escape from disciplines of validating theory. In the interests of the creative spirit, we need no longer worry over the accuracy of our evidence, or its (in)consistency with other work. But the bolder claim that theory can be considered grounded if discovered through research is rendered highly dubious if the research protocols through which theory is generated are relaxed. As Glaser and Strauss suggest, it may be highly desirable to generate social theory through the disciplines of research rather than from the comfort of the couch. But the disciplines of such research may be rather more demanding of researchers than is explicitly acknowledged in grounded theory.

CHAPTER 11

Grounding Grounded Theory

I began this book with the question: what is grounded theory? Now it is time to take stock: to summarize the various arguments we have considered and see where they have led us. We can begin with one of the central and recurrent tensions in grounded theory, that between theory generation and validation.

GENERATION AND VALIDATION

If we do try to combine the generation and validation of theory, this might have quite significant implications for the procedures of grounded theory. At the very least, its tolerant attitude to the (in-)accuracy of evidence—acceptable perhaps if no claims are made as to its veracity—might have to be abandoned. What would this involve? I suggested earlier that grounded theory is more concerned with processes than with people and places. To validate theory, we would have to consider its "external validity"—that is, whether or how far a theory of process applies to particular people or places. The systematic production of evidence might therefore require a different method of sampling and producing data, since the emphasis would shift toward obtaining data that is representative of people and places, and a more thorough and systematic study of each situation. Then the construction of categories (and the hypotheses connecting them) would have to be tested against all the available data—and partial categorization of data would no longer be adequate. Hypotheses connecting categories would have to be checked systematically against all the evidence. Thus the methodological shortcuts of theoretical sampling and theoretical saturation, which give grounded theory much of its exhilarating élan, would no longer be acceptable. This may seem to be too high a price to pay for the extra punch supplied by grounding theory while generating it, even given the development of software applications supporting qualitative analysis. While these applications can certainly facilitate a more rigorous analysis of qualitative data, they do little or nothing to ease the time and resource costs of its production.

The alternative, of generating theory without regard to its validation, is more compatible with some of the most appealing features of grounded theory, such as theoretical sampling and theoretical saturation. These are acceptable as procedures for generating theory, if not for validating it. The armchair may seem a still more attractive alternative, if we can distinguish (as I have suggested) the use of disciplined imagination—"imagination in a straightjacket"—from unbridled speculation in the generation of theory. This discipline for the imagination can be applied through conducting research, as Glaser and Strauss suggest, but it can also be supplied by other means. There may seem to be a paradox here, though only if we believe that the creative imagination should not be fettered. But it is surely the straitjacket imposed by prior knowledge that distinguishes the scientific imagination from more artistic and aesthetic forms of creativity.

STARTING FROM SCRATCH

This suggests that theory generation may be fostered rather than hampered by thorough immersion in one's discipline (how appropriate that term seems in this context) prior to research. In practice, even researchers explicitly adopting a grounded theory approach sometimes have difficulties resisting reference to prior theory (cf. Strauss & Corbin, 1997). Glaser and Strauss do encourage theoretical sensitivity through recourse to other resources, such as novels, personal experience, other disciplines, and so on, but they are most reluctant to don the conceptual straitjacket supplied by the discipline itself. They assume that such a straitjacket precludes a creative response to the data, which as a result can no longer yield its secrets freely, but is forced into some preconceived framework. We are therefore presented with a stark choice: between a "tabula rasa" approach, albeit tempered by some "theoretical sensitivity," and the unmitigated imposition of preconceived ideas on data. There is no hint here that conceptual frameworks can act as guides rather than as prison guards—that prior conceptions need not become preconceptions. As I have remarked elsewhere, there is a difference between an open mind and an empty head. In a similar vein, I also compared fixed and flexible ideas to a light cast by a lamppost and a torch; for when it comes examining data, the torch will certainly prove better than searching in the dark (Dey, 1993, p.229).

The equation of prior conceptions with preconceptions is evident in the language of "discovery" and "emergence," terms that suggest that observations can themselves force appropriate conceptualizations from us, as though the observer were not deeply implicated in the very act of observation. Even in physics, it has proved impossible to retain this separation of what is observed from how it is observed. This does not mean that we can observe whatever we like—or act as we please. The world is rather resistant to arbitrary action or conceptualization (so you cannot choose to observe human beings with 10 legs, for example). From an evolutionary perspective, our conceptualizations must be adaptive (on the whole—we may be wildly wrong in particulars) or we would not have survived. But it does mean that what we observe depends on our perceptions and actions as observers—where we go, what we do, what we look at, what we see, what we notice, how we interpret it, and so on—so that any observation is already subject to a series of conceptual filters. It would be needless to labor this point—since our experience makes it so obvious—were the language of discovery and emergence not so salient in grounded theory. It is perhaps this faith in grounded theory in the power of observation, apparently freed from all preconceptions, that sustains the discursive slide from generating theory to verifying it. What is observed becomes self-evident both to the observer and to those in the field, so that there is little added value to be gained from going through any process of further testing.

CATEGORIES, PROPERTIES, AND DIMENSIONS

The (implied) claim that observation can be preconceptual conforms to the way categories are defined and used in grounded theory. Categories are defined as concepts that stand by themselves as elements of a theory. This conceptual interpretation of categories is useful in distinguishing them from names or labels, which may lack any conceptual function. However, just what these conceptual elements are is not easy to establish.

This is partly because of the muddle between categories, properties and dimensions, as no rigorous distinction is drawn between the three concepts, which tend to be used interchangeably. Taxonomic and substantive relations between categories (and their properties) are confused. To clear up these muddles, I suggested that we make the following distinctions:

- *Category*—used as a way of identifying or distinguishing something based on comparison with other things.
- *Property*—used to ascribe a quality or attribute to something based on analyzing its interactions with other things.
- *Dimension*—used to measure extension.

If we are to make sense of categorization, then we have to distinguish clearly between categories (which are aspects of classification, assigned through comparison) and properties (which are attributes of things, analyzed through interaction) instead of using these terms interchangeably.

CONCEPTS AND INDICATORS

According to Glaser, the use of categories in grounded theory is based on a concept-indicator model. This analytic function does not seem entirely consistent with the sensitizing role of categories. Nevertheless, categories are based on indicators that convey "underlying uniformities" in the data, identified through comparison of similarities and differences. These indicators are combined conceptually (that is, in terms of their meanings) to generate categories, rather than through any mathematical operations. Glaser defines the meaning of a category in terms of its "operational meaning"—that is, not in terms of some abstract standard definition, but in terms of how it is specified in the data. These operational meanings are allowed to vary according to the context of the inquiry—there is no need for consistency in their application, as the object is to generate ideas rather than assess evidence.

I noted some problems raised by this reduction of concepts to a flexible range of indicators. First, it becomes difficult for concepts to play a general role in theory construction, if their meaning varies according to context. It is hard to see how concepts can "acquire a life of their own" independent of the evidence through which they are generated, if their meaning is "operational" rather than abstract. Second, it assumes that the logic of discovery overrides that of validation, despite claims to verify categories through theoretical saturation. In contrast to the apparent stress on flexibility, the underlying assumption in grounded theory is that we can recognize appropriate categories with sufficient consistency to ensure that categories are stable enough conceptually and empirically.

CATEGORIZATION RECONSIDERED

This in turn reflects a deeper issue about the use of categories in grounded theory—the implicit adoption (in the search for similarities and differences) of a classic model of classification. In the classic model, we have no problems in making distinctions, so things can be assigned unambiguously to different categories. A category is defined in terms of indicators shared by all members in common, but not by nonmembers. To generate a category, we have to identify those indicators that (taken together) are necessary and sufficient to confer membership of a category. These indicators provide us with rules which define membership unambiguously—so that all members and only members of the category satisfy these rules.

One problem with this view of classification is that it adopts a folk perspective that takes little or no account of the evidence on how we make distinctions. Here lies a danger for "naturalistic" forms of inquiry, which model themselves on common sense assumptions about how we think (or act)—which may prove misguided. I discussed a number of different accounts on how we form and use categories, all of which are critical, in one way or another, of the classic position.

The "prototype" account suggests that category membership is often assigned on the basis of family resemblance to a prominent exemplar, distinguished in terms of a cluster of ideal attributes that may bear little if any relation to any features common among other members. I might regard a study as an example of grounded theory on the basis of its similarity to work done by Glaser and Strauss (prototypical grounded theorists) rather than on the basis of meeting some rules specifying common features of grounded theory (as we have seen, these rules are likely to prove rather elusive). The prototype figures as a basis for membership through family resemblance rather than features common to all members. The use of exemplars drawn from memory offers a related but slightly different method of assigning membership. Here we can make judgments about membership based on a known example—rather than an "ideal" or "central" example. When we refer to "similarities and differences" as a basis for generating categories or assigning membership, therefore, we have to clarify just what it is we are making comparisons with. It may mean comparison with a previous exemplar or a prototypical ideal rather than a set of features shared in common.

We also need to think about how we make such comparisons. Even in a rule-based model, it seems that the best we can do is converge upon an approximation that discriminates among confusable alternatives. With recalled exemplars or ideal prototypes, we can make judgments in different ways: for example, by making a holistic judgment about resemblance or by taking a more analytic tack and identifying separate attributes for comparison. In the latter process, we can also use additive and multiplicative "processing," depending on whether separate attributes can be treated independently or they have a combined effect on outcomes.

The classic model implies that categories have firm boundaries such that something either is or is not a member. But this belies the way we make category judgments. With rules we are stuck with problem of contingency—for new alternatives may arrive that cannot be firmly allocated on the basis of existing rules. With family resemblance to an exemplar or prototype, paradoxically, we may be more confident about our judgments, precisely because we can ignore the problems associated with rules and boundaries. We judge on the basis of proximity to a familiar example. With "fuzzy logic" category boundaries are also uncertain, with the occasions where we can assign members unambiguously seen as special cases. For the most part, membership of categories is "graded"—that is, a matter of degree; some are seen as better (and some worse) members of categories than others. If membership is graded, it seems more difficult to define the boundaries of categories with any precision.

All these effects undermine classic classification, with its assumptions of sharp category boundaries and unambiguous membership assignations. But even more damaging to the classic account is the explanation Lakoff offers for these effects in terms of idealized cognitive models. The basic argument is that we do not make sense of categories by checking correspondence between them and some invariant features of the world. Instead, we make sense of categories in terms of underlying idealized cognitive models through which we invest (or "motivate," as Lakoff puts it) categories with meaning. Lakoff distinguishes a number of ways in which these models invest categories with meaning: including propositional rules, but also image-schematic structures and metaphorical and metonymic "mappings." Thus we categorize not through comparing similarities and differences to capture invariant features of things, but on the basis of cognitive models about what the world is like. I discussed examples such as "mother" and "bachelor," both categories which make sense not through correspondence with a world peopled by bachelors and mothers, but only in terms of (various and complex) underlying cognitive models of motherhood and bachelorhood. These cognitive models are idealized, animated by stereotypical assumptions and extended by conventional rather than logical connections.

These models are irreducibly cognitive, but they are also based on and informed by experience. Lakoff argues that our first categories are basic level expressing our bodily experience of the world, through which we grasp distinctions (in and out; over and under, etc.) that then become the stuff of metaphorical and metonymic projections of other categories. Therefore categories are stratified, with some more firmly rooted in (distinctively human) experience than others. Our "basic" categories are already internally structured—they are not "atomic" but already involve a holistic perception of experience in which the whole is more than the sum of the parts. Complex categories are not, therefore, to be understood as logical combinations or aggregations of simpler parts. Moreover, relationships between categories tend to be governed not by general rules but by conventions that have to be learned—they are culturally defined. The Dyirbal universe is an example of a conventional category structure that differs strikingly from our own—but is nevertheless rendered intelligible in terms of some conventional principles which shape the system.

Lakoff insists that categories are based on prior cognition, which is extended through various forms of conventional (rather than logical) reasoning. But even if we assign a central place in categorization to rules that discriminate among invariant features, it seems we have to acknowledge these conceptual underpinnings. Thus Harnad argues that all categorization must be approximate and provisional, as the rules we employ to discriminate among "confusable alternatives" involve selective attention to particular features and are always subject to revision in the light of experience. Moreover, he suggests that categorization involves "symbolic" as well as "iconic" forms of representation, and though language is more efficient it is also more abstract and less grounded in the particulars of perceptual representation.

As well as exploring methods of categorization and the meaning of categories, cognitive psychology examines the uses we make of them. It may make a difference, for example, whether we use a category for identification (what is this?) or discrimination (which of these?). We can also distinguish between the uses of categories for inference, instantiation, and combination. Inference from an exemplar to a category requires confidence that the attributes of the exemplar serve as reasonable cues as to category membership (high cue validity). Inference from a category to an exemplar requires confidence that members of the category will have a given attribute (high category validity).

These explorations of categorization take us far from the classic model of classification. They suggest that any straightforward generation of categories through identification of similarities and differences, as depicted in grounded theory, is misleading. If these various critiques of the classic model are accepted, then we need to be much more reflective about the way we generate categories and how we use them. We need to recognize the role of recalled exemplars and powerful prototypes in influencing categorical judgments. We need to examine more carefully the implications of permeable boundaries and graded membership. We need to identify the underlying cognitive models that invest categories with meaning. We need to consider the conventional links between categories, which use image-schematic, metaphorical, and metonymic "projections" as well as propositional schemas. We need to consider the kind of inference used in making categorical judgments.

This is no more than thinking about the tools of our trade rather than taking these for granted by accepting or simply assuming the standard model as prescribed by logic or common sense. But it does give a new twist to the claim that categorization involves conceptualization. It reinforces the earlier doubts I expressed about the possibilities of categorizing without prior conception. If categories derive their meaning from underlying cognitive models, it also queries the definition of categories as theoretical elements that somehow stand alone. It adds weight to the argument that we must think of categories in the plural as so many points on a string (or intersections of a web) through which we strive to develop our interpretation(s).

As I pointed out, most of this research on categorization has been done in the period since the initial launch of grounded theory. It is not surprising, therefore, that the classic model of classification informs the approach laid out in grounded theory. On the other hand, Glaser and Strauss place considerable emphasis on experience and practical knowledge as a basis for generating and confirming ideas. Thus grounded theory does not rely only on the classic model, and indeed anticipates the need for a more integrated approach to the issues of categorization. Subsequent research can therefore be seen as a way of making more explicit and consolidating strategies that already form part of a grounded theory perspective.

CODING

We can already recognize the force of Lakoff's arguments in the way the concept of coding has come to symbolize the central processes of qualitative analysis. Once we recognize the metaphorical and metonymic extensions that invest the concept with meaning, we acquire a clearer idea of the cognitive model underpinning coding as an analytic process. The metonymic extension takes a trivial, mechanical aspect of data processing in survey research—the assignation of numbers in place of categories for easy processing by computer—and extends this to the whole business of categorization. The metaphorical extensions from legal, linguistic or machine codes invest that process with overtones of consistent, logical, and mechanical translation from one context to another. Coding refers to the translation of data from its raw state (as newly produced transcripts, etc.) to a new state in which it is ready for or amenable to analysis. The "ready for" perspective assumes that data are first coded and then analyzed, while the "amenable to" perspective allows that analysis may occur as an aspect of coding. It is not always clear to which position coders subscribe—sometimes, it seems, to both. (Incidentally, I suspect that if we described those who advocate coding as coders rather than analysts, the term might soon disappear!) But in either case, the conceptual aspects of this translation process are obscured by the mechanistic overtones I have noted. That said, coding in grounded theory is said to proceed simultaneously with analysis and indeed is sometimes equated with it.

Glaser and Strauss present coding within the logic of discovery. They explicitly reject the idea that coding should precede analysis, since coding is conceived precisely as a method of generating theory. They are equally emphatic in rejecting the use of coding to quantify data in order to test hypotheses. So coding is an activity designed to generate theory systematically rather than to accumulate evidence. I suggested we can think of this approach as partial coding—just enough for the generation of theory and no more.

Glaser and Strauss describe categories produced through partial coding as theoretically saturated, but this, in my view, is another unfortunate metaphor. The claim that categories are saturated does not square with a procedure which stops (far) short of coding all the data. The cessation of coding is justified by a claim that no new properties are being generated; but it is impossible to validate this conjecture without actually coding all the data. Such categories might be better described as "suggested" by the data—when coding reaches theoretical sufficiency rather than saturation. I also suggested that the lack of fresh insights from the data may in any case reflect the increasingly focused and directive methods of sampling and data collection.

The approach to coding in grounded theory has been cited with approval by those concerned with the use of coding in the context of software development. The creative tasks of coding in generating theory are contrasted with its mundane functions in referencing text for retrieval. We saw that this contrast in the uses of coding takes a series of forms: referential versus factual (Richards & Richards, 1995); representational versus heuristic (Seidel & Ulle, 1995); and data reduction versus data complication (Coffey & Atkinson, 1996). The heuristic or data complication functions come closest to that of theory generation as envisaged in grounded theory, since they emphasize the discovery and conceptualization. The data reduction, representational, referential and factual functions concern the use of coding for data retrieval. Here coding is used to denote either the content or the location of data (or other case characteristics). Denoting content involves some reference to what is in the data, whereas denoting location merely refers to where to find the data. Thus coding can have three distinctive (if overlapping) uses: crudely, telling us where to find the data, what is in it, and what to do with it.

There are a number of dangers that are associated with confusing these different activities. One is that of reification, mistaking the codes for the data (and indeed, for the phenomena) to which they refer. Another is of fragmenting data—of losing more meaning by isolating segments of data from their immediate context than is gained by recontextualizing them for comparison through coding. More generally, there is a concern to distinguish (and protect) the heuristic role of coding from its (atheoretical) uses to denote content and location.

I suggested that these contrasts may be rather overdrawn, as no method of denoting content or even location is free from conceptual assumptions. Even the activity of indexing is not aconceptual, as any librarian will testify. An apparently mechanical procedure, such as numbering all lines of text for identification purposes, actually involves conceptualizing text in terms of lines rather than words, sentences, paragraphs or user-defined units of analysis. Categorization may proceed at different levels, with different purposes and degrees of flexibility (we may want to revise our categories more often than our units of analysis) but it remains nonetheless a method of conceptualizing the data. However, this point is obscured by terms such as "indexing" and "coding," neither of which begin to express the analytic nature of the tasks involved in the organization of data.

I suggested that, so long as we conceive of these activities as forms of coding, confusion will continue to reign. We should find appropriate terms for the different tasks involved. Much of the analytic work of identifying what is in the data and what to do with it may involve categorizing data. But categorizing does not exhaust analytic possibilities, particularly if we want to analyze substantive as well as formal relations. We need to think about ways of linking and connecting, and not just comparing. These would be better conceived not as ways of coding data but simply as aspects of analysis.

Let us now consider the choice between what I have called partial and full coding. The latter requires systematic coding of all the available data, while the former requires coding only to reach theoretical sufficiency. These options underpin the above discussion of the representational and heuristic functions of coding. However, we can recognize at least two further options. One is no coding and another is flexible coding.

The no coding option involves analyzing data without using an approach based on categorization. We might, for example, prefer to identify key narrative sequences or structures in the text. We might opt for a causal analysis, singling out key events and identifying the configuration of conditions and consequences in which they are embedded. We might opt for summary descriptions of key passages or develop profiles that connect various strands of evidence. None of these analytic methods requires a systematic search for similarities and differences in the data. Although Glaser and Strauss seem to dismiss a no coding option as inevitably "impressionistic," given the complex nature of conceptualization, and the rich and routine understandings that can inform it, it is less clear that we should privilege comparison as a more systematic approach.

The flexible coding option represents a halfway house between partial and full coding. I argued that, even within the logic of discovery, there may be a role for coding beyond the point of theoretical sufficiency. To this end, I took the rather unfashionable view that qualitative research can benefit from a mathematical approach. I defended this position partly by reference to the endemic role of numeracy in our everyday lives, which is underestimated by our common fears of and prejudice against mathematics. I also suggested that meaning and number are more inextricably bound together than is often imagined. But my main arguments concerned the utility of numbers in discerning patterns, considered as either distributional regularities or as underlying rules.

When it comes to discerning distributional patterns in the data, this requires recognition of regularities or repetitions, which implies a numerical dimension. We can use numerical data to capture and summarize these patterns, most conveniently perhaps translated into a graphical format. Mathematics may also prove useful when it comes to discerning patterns conceived as the underlying rules which, when allied with contingent events, may produce complex but nonrepetitive patterns (such as the Penrose tile). To take advantage of this potential resource, however, we may need to go some way beyond partial coding. Thus mathematics, even of a very limited kind, may help to generate theory, even setting aside any role it may have associated with the logic of validation.

Open Coding

As well as discussing when to stop coding Glaser and Strauss advise on how to start it—and on how to proceed. Here we encounter the well-known divisions between open, axiel, and selective coding as three distinct aspects of analysis. We open (and open out) the analysis with open coding, which involves breaking data down into discrete parts that are later stitched together again through theoretical connections. This fragmentation of data can be considered part and parcel of an analytic process, that requires resolution of something into simpler elements to see how they combine. There are two points to make about this procedure.

One is that we may not so easily separate one part from another, as though the identification of each part proceeds in splendid isolation; any more than I can separate out parts of a chair (seat, back, legs) without already understanding something of the relationship of each part to the other—and to the whole. If we consider parts in isolation—as unrelated—then we may have a much harder time putting them together again. Therefore an analytic approach need not imply the temporal separation: first identify your parts and then combine them. It may be more productive to consider these as two sides of a single coin. Moreover, the coin we use is part of the currency with which we work. People who make chairs are more likely than people who just sit in them to identify other parts (such as joints).

The second point is based in part on my earlier argument that the whole is more than the sum of its parts—just as a chair is more a back, seat and legs, for these parts have to be related in a particular order and located in a particular position. It also inspired in part by Kosko's claim that the whole is "in" the part, just as the seat of a chair would not be a seat were it not for the chair. This suggests that we should not think of holistic accounts as separate from analytic ones. The whole picture informs our understanding of the parts and their connection, just as our understanding of the chair as a whole informs our identification of its various parts. Holistic views are often dismissed (and I confessed myself as a culprit here) as a result of "mere" impressions and intuitions. But holistic knowledge is not so easily discounted. I know how a chair works because I sit on it. I suggested that this experiential knowledge informs my holistic understandings of things like chairs, shopping, banks, and, indeed, whole tranches of social life. I also suggested that this experiential knowledge could be both rich—because as human beings we inherit a world in which knowledge has accumulated; and routine—because it is experienced and understood through everyday activity.

This experiential knowledge may provide a useful starting point for developing theory. In open coding we are encouraged to engage in a line-by-line analysis, which is seen as the most generative form of coding. The coding of wider units is presented as a way of facilitating a fine-grained analysis. However, this may not be the most productive approach if it inhibits the identification of how parts relate to each other and to the whole.

Axiel Coding and Theoretical Codes

In grounded theory the data fragmented by open coding is put together again through what Strauss and Corbin call axiel coding. In their version of grounded theory, this involves the application of a coding paradigm, which we considered in more detail in relation to the analysis of process. In his critique, Glaser argues that the coding paradigm is only one of a family of theoretical codes through which categories can be connected. These families form a rather disparate crew, and some seem to bear little relation to coding per se. However, Glaser's critique does raise some interesting issues. Foremost among these is how the identification of such preconceived codes can be reconciled with the usual insistence on emergence. The disparity is dealt with by suggesting that the use of a theoretical code must reflect its emergence in the data. The very plurality of theoretical codes safeguards against imposing a particular ("pet") coding family on the data—which perhaps explains some of Glaser's hostility to the idea of a coding paradigm. This strategy may not be altogether successful, however, since it leaves unclear how empirical cues can be recognized (without already using a coding family). For example, are stages indicated by the data or a construct applied by the researcher to order events? Glaser's distinction between substantive and theoretical coding implies the former, but he also suggests that theoretical coding is implicit in substantive coding, implying the latter.

Selective Coding

Finally, selective coding involves the identification of a core category around which the inquiry then revolves. This serves as a way of both delimiting and integrating theory. The delimitation is necessary because of the density of grounded theory—that is, to avoid both researcher and reader being overwhelmed. The integration is necessary for clarity and coherence. Although core categories are discovered, these are general utilitarian considerations rather than dictated by what emerges from the data. Indeed if there are several promising candidates for a core category, its selection can be quite arbitrary. This privileging of one core category as a fulcrum for analysis may therefore be quite misleading, giving an unwarranted emphasis to its explanatory powers (reflected in its demotion in the next study). The insistence on a single analytic fulcrum may distort relationships in which no one category is central. Also, the use of a core category may encourage a one-dimensional analysis, which presents a simple perspective but at the expense of competing explanations.

PROCESS

Sandwiched between open coding and selective coding is what Strauss and Corbin call axiel coding via the coding paradigm. One reason they insist on this is that the analysis of process becomes an indispensable requirement in their version of grounded theory. This involves (Glaser would say, "imposes") the use of an analytic framework composed of conditions, interactions, and consequences—with consequences in turn conceived as creating new conditions (and hence new strategies) for interaction. Process is considered as "the linking of action/interactional sequences, as they evolve over time." As well as relating conditions to consequences through interaction, Strauss and Corbin analyze various aspects of process such as the pace, degree and direction of change. Process is thus conceived as involving change over time, but Strauss and Corbin (though not Glaser) also consider as process those interactions which lead nowhere in particular, such as constant adjustment to maintain a status quo.

Strauss and Corbin also propose a method of analyzing process through a "conditional matrix" and the "tracing conditional paths." The conditional matrix sets out the conditions in which interaction is embedded according to their level of generality. The conditional path traces out the way that specific conditions produce consequences through their effects on interaction. In tracing these paths, Strauss and Corbin distinguish between contextual, causal, and intervening conditions, though they do not conform to standard distinctions between types of conditions. To Glaser these prescriptions are anathema, since they not only privilege the coding paradigm (ahead of other theoretical codes) but shift the focus of analysis from identifying and explaining observed variations through constant comparison to what he calls "full conceptual description."

In the conditional matrix, Strauss and Corbin stress differences in the generality of conditions. However, there are other distinctions among conditions that we may also want to take on board. We can distinguish whether conditions are necessary or sufficient to produce a particular effect. If necessary, a condition must be present for the effect to occur; if sufficient, no other condition need be present for the effect to occur. Only where a condition is both necessary *and* sufficient do we consider it as a causal rather than contextual condition. We can distinguish a single condition (where we can attribute an effect to a particular factor) from multiple or conjunctural conditions (where an effect results from different factors, or the interplay of more than one factor). We can also distinguish local conditions (which explain a single event in terms of a specific combination of conditions) from general conditions (which explain events in terms of generalizations). I also suggested that we can distinguish contextual, causal, and intervening conditions more effectively if we take account of temporal sequence. In brief, contextual conditions cannot follow causal conditions in a temporal sequence, while intervening conditions must do so. Failure to draw these distinctions leads a rather confusing account of process in which it is not clear how we can distinguish causes from context; we can confuse local explanations with generalizations; and temporal sequence can become an incidental rather than an integral feature of analysis.

CAUSALITY

To understand process, I suggested, we have to distinguish clearly between correlation and causality. Our understanding of causality is rooted in the intelligibility of connections between cause and effect—that is, in our understanding of how things work. I suggested that this is rooted in our bodily experience of the world—we ourselves make things work; but also in more abstract (scientific) accounts of how causes can produce effects.

This emphasis on intelligibility contrasts with the analysis of causality in terms of constant conjunction. Following Ragin, we considered two ways of inferring causality from constant conjunction: the methods of agreement and difference. The method of agreement depends on observing an invariant pattern—that whenever we observe a putative effect, we also observe the putative cause. The method of difference depends on observing variation—that whenever we do not observe the putative cause, the putative effect is also absent. Neither of these methods of inference, however, can cope well with multiple or conjunctural causation.

Glaser and Strauss explicitly reject analytic induction, which tries to identify causality through the method of agreement (that is, by looking for constant conjunction and rejecting hypotheses in the event of negative instances). Glaser and Strauss appeal to the logic of discovery to exempt grounded theory from concerns with verification—so the negative instance becomes incorporated as an example of further variation. They also endorse a multicausal view of causality, rejecting any attempt to isolate the effects of single factors across a range of situations. Clearly, claims about causal inference in grounded theory are not based explicitly on the methods of agreement and difference. Therefore there seems to be some inconsistency between the language of grounded theory—using similarities and differences to identify regularities in the data and so on—and its underlying logic.

One clue as to how Glaser and Strauss may think causal relations can be identified, is the remark that "general relations are often discovered *in vivo;* that is, the field worker literally sees them occur" (Glaser & Strauss, 1967, p.40). While this remark seems to imply that observation can be atheoretical, it also suggests that causality can be distinguished on a different basis than constant conjunction. Following Sayer, we looked at the suggestion that causality can be analyzed in terms of powers (of things) rather than patterns (of events). That is, causality is inherent in powers (or properties) of things—such as an employer's power to hire and fire labor. We can identify such powers by investigating the "structure and constitution of the objects which possess them." We can identify what it is possible (or impossible) to accomplish, given a particular structure. In this way, we can render causality intelligible in terms of how things work.

STRUCTURE AND AGENCY

Although there is some reference to social structure in grounded theory, this tends to be only as a precondition or a by-product of interaction. The analysis of structure is most explicit in Glaser's work on theoretical codes where structural processes can themselves figure as the object of inquiry. However, Glaser restricts the analysis of structural change by insisting that it must be exhibited in stages relevant to changes in interaction; and further by interpreting stages only in terms of organizational expansion and contraction. At least Glaser's insistence on stages incorporates a temporal dimension, if unreasonably restricted. In their conditional matrix and tracing of conditional paths Strauss and Corbin seem content with an atemporal, cross-sectional analysis of current conditions that does not capture the evolution of events over time.

Following Archer, I considered ways of incorporating temporality more effectively into the analysis of process. Like Sayer, Archer also focuses on how structures can act as generative mechanisms (that is, exercise causal powers).

Like Sayer, she analyzes these mechanisms in terms of the internal relations and resulting capacities of emergent structures. This is in some measure a holistic view—structures emerge with complexities and capacities which cannot be attributed to individual agents. It also requires a temporal perspective, as emergent structures both result from earlier interaction and pre-exist current interaction. This is consistent with the temporal sequence presented in the coding paradigm. But Archer insists that structure and agency should not be elided; whereas in grounded theory, process is conceived as a sequence of interactions, but structural conditions are recognized only through their (contemporaneous) impact on interaction rather than as causally efficacious in their own right.

I used Archer's own work to serve as an example of how such an analysis can proceed. She distinguishes between different kinds of structural capacity—for example, properties with some material basis (factories, schools) from those that are entirely cultural (such as recipes). These have different kinds of "internal relations" with different implications for interaction. Archer also distinguishes among "persons," "actors," and "agents," analyzing these in terms of different forms of agency. For example, agents are analyzed in terms of opportunity structure and its implications for the identification and pursuit of vested interests, and ultimately for the reproduction or transformation of social structures. Archer also discusses the latter in terms of the situational logics associated with different structural patterns—for example, whether or not structural relations are mutually reinforcing or contradictory and whether opportunities arise that tend to disrupt these relations, or foster them.

While we can question the detail of Archer's analysis, it does suggest some useful ways of exploring the impact of structural conditions on interaction. We can ask how far structural relations are inherent or contingent and whether they are complementary or contradictory. We can consider different situational logics and whether these encourage different strategies, such as compromise or conflict. We can analyze interaction in terms of a differentiated view of people—as persons (and their needs); actors (and their roles), and agents (and their interests). We can consider the implications of interaction for "structural elaboration" in terms of relations of power and exchange. These various suggestions are not immune from criticism, but at least they do begin to address more specifically the question of how process can be analyzed.

THEORIZING

In terms of theorizing, the main distinction Glaser and Strauss make is between substantive and formal theory. This, I suggested, is largely conceived in terms of levels of abstraction, which provides a rather weak basis for discriminating between the two. An alternative account that contrasts empirical and conceptual theory is no more satisfactory. A further distinction, in terms of whether we study a particular situation, or a range of situations, is more promising, though only if we specify more closely the differences (and overlaps) between theory that explains the particular and theory that produces generalizations.

I suggested that studies of particulars must locate them in time and space. This allows us to specify a phenomenon uniquely, since we can safely assume that no other event can occupy the same space and time. This is not really possible within grounded theory, even of a substantive kind, since even here the method of comparative analysis requires abstraction from time and space. The objects of inquiry are not cases with particular identities but situations defined only in terms of similarity and difference. This perhaps explains why Glaser and Strauss are obliged to distinguish substantive and formal theory in terms of different levels of generality.

But the level of generality becomes an irrelevant consideration if we define a substantive theory as one that accounts for a particular phenomenon, defined in temporal and spatial terms, and a formal theory as one that aims to produce generalizations. The particular phenomenon we study may be more or less general, if by this we mean it can be narrowly or broadly conceived. We can study anything from the latest fashion fad to the evolution of humankind. The same applies to formal theory, for we can produce generalizations with a narrow or a broad application—anything from what happens when people watch too much television to the effects of ecological concerns on human behavior.

Since both substantive and formal theory in grounded theory are intended to produce generalizations through comparative inquiry, to avoid confusion we can follow Wallerstein in describing theory explaining particulars as idiographic and theory producing generalizations as nomothetic. The former uses intensive methods of inquiry, orientated to the analysis of particular conjunctures, multiple and conjunctural causality, and a holistic and evolutionary account of events (often in narrative form). Nomothetic theorizing abstracts from particular contexts to produce general explanations, typically by isolating key factors and their effects across a range of cases. It usually proceeds through propositional logic (if A then B) rather than narrative form. Variables are regarded as stable and independent across space and time rather than defined a particular context.

Although it seems more orientated to producing nomothetic generalizations through both substantive and formal theory, grounded theory mixes both intensive and extensive methods of inquiry. The intensive approach is evident in its qualitative orientation, in its analysis of process, and in the professed interest in narrative accounts and the production of complex and "dense" theory. The extensive approach is evident in the analysis of variables, the language (if not the logic) of analyzing causality in terms of regularities, the selective focus on a core category, and the orientation to generalization. It is not clear, though, whether in mixing these modes of theorizing grounded theory offers the best or the worst of both worlds. From an intensive standpoint, the loss of intimate knowledge of a particular phenomenon may outweigh the gains of comparison between many situations. From an extensive standpoint, the dilution of comparative logic as a basis for causal inference may outweigh the virtues of recognizing complex forms of multiple and conjunctural causality.

Nevertheless, one great attraction of grounded theory undoubtedly lies in its mix of methods. The combination of intensive and extensive methods of theorizing takes us into new territory, through a more dialectic and dynamic process of moving between intensive focus on particular sites and extensive theorizing based on comparison. This approach tries to harness useful features of each mode of theorizing. I suggested that the ambition to mix methods may be encouraged by recognition that the two modes of theorizing are not after all so conflicting. Idiographic theorizing involves nomothetic generalizations, and nomethetic theorizing has to apply these generalizations with given spatial and temporal contexts. Idiographic theorizing may aim to explain the particular, but that does not exclude the use of generalizations (even though it may not aim to test them). Indeed, I argued that we can hardly hope to identify any particular event, let alone explain it, without relying on generalizations. On the other hand, nomothetic theorizing may aim to produce generalizations, but these can only apply to events within a delimited context. The latter may be rarely specified, but that does not make this constraint any less stringent. Instead of presenting these as antithetical perspectives, therefore, we ca help to ground grounded theory by recognizing them as two sides of the same coin (theorizing), though differing in their focus and emphasis.

I reinforced this eclectic view of idiographic and nomothetic theorizing by suggesting that the criticisms that Glaser and Strauss advance against each form of theorizing are rather misplaced.

I defended nomothetic theorizing against their criticisms of speculation and logical deduction, suggesting that these confuse a variety of different targets, some legitimate, but others not. One of the main planks of grounded theory is the generation of theory through the discipline of conducting research. I suggested that there are other disciplines—in the form of existing evidence and theory—that can also be used as a "straitjacket" for the theoretical imagination.

I also defended idiographic theorizing against the criticisms that Glaser and Strauss make of the individual "case study." Whether as an "encased" study or a "case of" study, this can be used effectively to develop theory—so long as we abandon the (needless) insistence on generation of theory through direct observation and empirical comparison of "similarities and differences."

The encased study, for example, allows us to examine how things work—things which may be unique (such as the Internet) but of vital importance to our present circumstances.

VALIDATION

This leads us to the question of what theory is for—and how such aims can be realized. Glaser and Strauss distinguish a range of objectives, the most notable of which concern the prediction of behavior and (dependent on this) the production of practical applications: theory that both fits and works. The claim that theory fits the objects of study might seem sustainable only if theory is validated through the process whereby it is generated. But Glaser and Strauss also suggest a theory can be validated through its practical application. This leaves some work to be accomplished by those applying the theory, for they have to assess its relevance to the situation and adapt it accordingly. Although this therefore involves a process of continual adjustment, theory has still to retain both sufficient generality to apply across a range of contexts and nevertheless convey a holistic view of the total picture.

I suggested that this may in fact prove an impossible burden, even for theorists let alone practitioners. The ability to make predictions depends on being able to specify the total picture, but this is possible only within closed systems. In the more or less open systems of society (which reflect our capacity to initiate change) it is hard to extend explanation of how things work into predictions of what will happen, even if we can specify initial conditions. (Perhaps the best we can hope for is to recognize a range of (im-)possibilities that can then provide guides to action.) I suggested this range of possibilities lies between the unpredictable and the arbitrary and that it reflects the degree to which structures are entrenched or open to transformation. We can differentiate between possibilities in terms of various criteria, including the timescale associated with different kinds of change and our understanding of processes and our confidence in that understanding. From an evolutionary perspective, we can consider the extent to which adaptations continue to fit a changing environment, and also whether people become locked in to particular processes or can break out in new directions. All this can confer some control over events—in the sense of providing a guide to action. In this sense, it is not necessary to predict events in order to control them.

We might regard such control (as against prediction) as an adequate test of the practical adequacy of a theory. Glaser and Strauss flirt with this view, though it is not consistent with one of the other aims they attribute to theory—that of "theoretical advance." That hardly matters for substantive theory, as they argue that the field tends to move on anyway, before any more rigorous testing is possible. But with formal theory, generalized to many different contexts, we surely need something more. We should distinguish, I suggested, between the utility and validity of a theory. Whereas a claim to utility depends upon one's interests, a claim to validity has to be conceded (or denied) regardless of one's interests. We still have to consider the grounds upon which a theory can claim validity, without reducing this to the question of whether it is useful in some practical context.

I defined validity in terms of being "well-grounded conceptually and empirically." The claim of grounded theory to produce a valid account conforms to this prescription insofar as the concepts it generates are grounded in the data it produces. But there are more stringent disciplines to be met, that may be compromised somewhat by the grounded theory approach. On the conceptual side, the rejection of preconceived ideas may encourage unfettered creativity (or reinventing the wheel) but it downplays the disciplined search for consistency or continuity between new and established ideas. This may be somewhat mitigated by the role of the theorist in bringing theoretical sensitivity to the task of interpretation, but this merely raises the question of why some prior conceptions are allowable but a more systematic appraisal of current theory is not. Conceptual validity also requires clarity in the use of arguments, which is difficult to reconcile with the deliberate loosening of logical procedures in grounded theory. And it requires the critical assessment of alternative accounts rather than their collapse into a unidimensional core category.

On the empirical side, grounded theory at least locates theory generation within the disciplines of empirical research. But it does so only within the confines of the current project, rather than through a confrontation with evidence produced through earlier research. Finally, the procedures such as those for sampling and analyzing data (theoretical sampling and theoretical saturation) may be acceptable for generating theory but they seem of more doubtful value for grounding it.

None of these considerations invalidates efforts to produce grounded theory; indeed, if anything, they encourage it by recognizing the virtues of generating theory through the disciplines of confronting evidence. Though I defended speculation against the strictures of Glaser and Strauss, it was speculation "in a straitjacket" and not as a license for free and fanciful theorizing. The basic impulse of grounded theory—to generate theory through confrontation with evidence—can be honored even if the disciplines required for validation are recognized as more demanding than it allows.

Much the same point applies elsewhere. While the concept of categories employed in grounded theory may be inadequate, recognition of the complexities of categorization can only strengthen grounded theory. While the procedures advocated in grounded theory for coding data are contestable, recognition of the importance of holistic and substantive connections can complement the contribution of constant comparison to the generation of categories. Though the analysis of process in grounded theory may be limited, it can be strengthened by the recognition of emergent properties and the dynamic interplay of structure and agency over time. Though the distinction between substantive and formal theory in grounded theory may be rather imprecise, recognition of the respective roles of idiographic and nomothetic theorizing can provide a clearer context for the use of intensive and extensive methods in grounded theory.

It might be argued that any theorizing that accepts such points as these will no longer remotely resemble grounded theory. However, Strauss and Corbin took a more accommodating view. Asked what features of the methodology are central, they suggested the following:

> . . . the grounding of theory upon data through data-theory interplay, the making of constant comparisons, the asking of theoretically oriented questions, theoretical coding, and the development of theory (Strauss & Corbin, 1995, p.283).

This is indeed a broad umbrella and one which allows for evolution in any number of directions. But in any case, surely nothing could be more arid than to waste energy in disputation over a name. As Lakoff might argue, what does matter is the underlying cognitive model that invests that name with significance and how reflection on that model can help us recognize—if not resolve—the problems that confront us in trying to conduct research.

Epilogue

I began this book with a puzzle prompted by the relationship of grounded theory to software applications (my own included) for analyzing qualitative data. The advent of computer assistance for qualitative data analysis has encouraged diffusion of a methodology widely seen—by developers among others (Lonkila, 1995)—to have an affinity with software because of its emphasis on coding data, systematic procedures and rigorous analyses:

> The widespread influence of computer-assisted qualitative data analysis is promoting convergence on a uniform mode of data analysis and representation (often justified with reference to grounded theory) (Coffey et al., 1996, abstract).

These authors argue that the centrality of coding in both software for qualitative data analysis and in grounded theory promotes an "unnecessarily close equation of grounded theory, coding, and software" (Coffey *et al.*, 1996, p.7.5):

> There is, therefore, a danger that researchers may be led implicitly toward the uncritical adoption of a particular set of strategies as a consequence of adopting computer-aided analysis (Coffey et al., 1996, p.7.4).

Both these arguments—that software has promoted a convergence of methods around grounded theory and has fostered an uncritical attitude to methodology—have been contested. Lee and Fielding (1996, p.2.3) argue that we need to take account of the diverse ways in which researchers use coded data. They also note that "developers popularizers and commentators have often stressed the need for epistemological awareness and debate in relation to software use" (Lee & Fielding, 1996, p.2.2). However, there can be little doubt that software developments have generated interest in and added impetus to the diffusion of a grounded theory approach. This is evident not so much in the following assertion as in the very fact that the authors felt obliged to make it:

> We should neither assume that qualitative research only involves grounded theory nor that CAQDAS [computer-assisted qualitative data analysis] supports only a grounded theory approach. There are other viable approaches to qualitative analysis, and indeed the practice of qualitative researchers may be rigorous without their explicitly employing any of the several approaches blessed with a memorable "label" and a school of avowed followers (Lee & Fielding, 1996, p.3.1).

Acknowledging the influence of software development, Strauss and Corbin could report with a mixture of satisfaction and apprehension that

> the diffusion of grounded theory procedures has now also reached subspecialties of disciplines in which we would not have anticipated their use—and does not always appear in ways that other grounded theorists would recognize as "grounded theory" (Strauss & Corbin, 1994, p.277).

As the authors (modestly) comment, "the methodology now runs the risk of becoming fashionable" with users claiming to use the method without understanding its basic tenets. Lee and Fielding likewise note that

> When qualitative researchers are challenged to describe their approach, reference to "grounded theory" has the highest recognition value. But the very looseness and variety of researchers" schooling in the approach that the tag may well mean something different to each researcher. (Lee & Fielding, 1996, p.3.1).

They suggest that the grounded theory tag may often be claimed for purposes of legitimation—being associated with "a memorable 'label' and a school of avowed followers"—despite differences between the methodology adopted and that put forward either in the original text or subsequent reformulations.

In the same vein, Bryman and Burgess suggest that grounded theory is "widely cited as a prominent framework for the analysis of qualitative data" (1994, p.220), though they question how far it is directly applied. The influence of grounded theory is most apparent, they suggest, in an appreciation of the "desirability of extracting concepts and theory" out of data, and in coding and "the use of different types of codes and their role in concept creation" (Bryman & Burgess, 1994, p.220).

It seems that anxieties over the convergence of qualitative research around a single methodology, which takes coding as the core of theorizing, may be well-founded. The prominence given to coding in grounded theory lends legitimacy to the use of software applications to code data and then to generate theory by relating categories through various forms of retrieval. This use of software to code and retrieve data has attracted criticism not least from those responsible for developing software to assist in qualitative analysis. However, such criticisms tend to endorse the logic of discovery offered by grounded theory without abandoning its claims to validation. Indeed, the introduction of software has been accompanied by arguments for greater rigor in qualitative research; and the capacity of the computer to speed up mechanical processes (such as assigning categories or retrieving data) and handle a larger volume of material has encouraged a stronger commitment to validation.

Pleas to recognize the heuristic or complicative functions of categories are liable to fall on deaf ears, however, so long as the basic framework of a grounded theory approach is taken for granted. The inclination to consider coding as an aconceptual process is not just a product of the technological capacities of the software. It is rooted in a grounded theory methodology that at times seems to present observation as atheoretical, coding as emergent rather than constructed, and theory as something we "discover." It is reinforced by the conception of categories as separate concepts that are later connected, thus downplaying more integrative and holistic conceptions of how analysis proceeds. It is further reinforced by the analysis of process through cross-sectional comparisons of "slices of time" rather than through an evolutionary analysis of emergent powers and the interplay of structure and agency.

In other words, complaints directed against the misuse of software may be misdirected if the seeds of an overly mechanistic approach have already been sown by the underlying methodology with which the software is often associated. This would certainly be a striking example of unintended consequences, given the stress in grounded theory on creativity, conceptualization, and theoretical endeavor. However, people, like water, tend to follow the easiest course, and the danger of the new software technology is that it sets one such course for us (interpreting grounded theory largely in terms of coding) while diverting our attention from other possibilities. These possibilities can be detected in the grounded theory methodology, though they are sometimes implicit rather than explicit. The dangers of a mechanistic approach cannot be avoided merely through exhortation. It requires reflection on the origins of the particular path taken and the problems it leaves unresolved. If this book has contributed to that end for others as well as myself, then it will have served its purpose.

Bibliography

Andreski, S. (1974). *Social science as sorcery*. Harmondsworth, UK: Penguin.

Altheide, D. L., & Johnson, J. M. (1994). Criteria for assessing interpretive validity in qualitative research. In N. K. Denzin, & Y. S. Lincoln (Eds.), *Handbook of Qualitative Research*. London: Sage.

Annells, M. (1996). Grounded theory method—Philosophical-perspectives, paradigm of inquiry, and postmodernism. *Qualitative Health Research, 6*(3), 379–393.

Archer, M. S. (1995). *Realist social theory: The morphogenetic approach*. Cambridge: Cambridge University Press.

Backett, K. C. (1983). *Mothers and fathers: a study of the development and negotiation of parental behavior*. Edinburgh: Edinburgh University Press.

Baker, C., Wuest, J., & Stern, P.N. (1992). Method slurring: The grounded theory/phenomenology example. *Journal of Advanced Nursing, 17*, 1355–1360.

Barclay L., Everitt L., Rogan F., Schmied V., & Wyllie A. (1997). Becoming a mother—An analysis of women's experience of early motherhood. *Journal Of Advanced Nursing, 25*(4), 719–728.

Barrow, J. D. (1993). *Pi in the sky: Counting, thinking and being*. London: Penguin.

Barrow, J. D. (1997). *The artful universe: The cosmic source of human creativity*. London: Penguin.

Benoliel, J. Q. (1996). Grounded theory and nursing knowledge. *Qualitative Health Research, 6*(3) 406–428.

Bohm, D. (1983). *Wholeness and the implicate order*. London: Routledge.

Brannen, J., Dodd, K., Oakley, A., & Storey, P. (1994). *Young people, health and family life*. Buckingham, UK: Open University Press.

Bryman, A., & Burgess, R. G. (Eds). (1994). *Analyzing qualitative data*. London: Routledge.

Charmaz, K. (1990). Discovering chronic illness: Using grounded theory. *Social Science and Medicine, 30,* 1161–1172.

Coffey, A., & Atkinson, P. (1996). *Making sense of qualitative data: Complementary research strategies.* Thousand Oaks, CA: Sage.

Coffey, A., Holbrook, B., & Atkinson, P. (1996). Qualitative Data Analysis: Technologies and Representations. *Sociological Research Online, 1* (1).

Creswell, J. W. (1998). *Qualitative inquiry and research design: Choosing among five traditions.* London: Sage.

Darling, D. (1989). *Deep time.* London: Bantham Press.

Davis, J. (1994) Social creativity. In C. M. Hann (Ed.), *When history accelerates: Essays on rapid social change, complexity, and creativity.* London : Athlone Press.

Denzin, N. K., & Lincoln, Y. S. (1994). Introduction: Entering the field of qualitative research. In N. K. Denzin, & Y. S. Lincoln (Eds.), *Handbook of qualitative research.* London: Sage.

Dey, I. (1993). *Qualitative Data Analysis: A user-friendly guide for social scientists.* London: Routledge.

Diamond, J. (1992). *The Rise and Fall of the Third Chimpanzee.* London: Vintage.

Dildy, S. P. (1996). Suffering in people with rheumatoid arthritis. *Applied Nursing Research, 9*(4), 177–183.

Dunbar, R. (1997). *Grooming, gossip and the evolution of language.* London: Faber & Faber.

Eve, R. A., Horsfall, S. & Lee, M. E. (Eds.), (1997). *Chaos, complexity and sociology: myths, models and theories.* London: Sage.

Fielding, N. G., & Lee, R. M. (Eds.), (1991). *Using computers in qualitative research.* London: Sage.

Fischer, M. D. (1994). Modelling Complexity and Change: Social Knowledge and Social Process. In C. M. Hann (Ed.), *When history accelerates: Essays on rapid social change, complexity, and creativity.* London : Athlone Press.

Fisher, M. (1997). *Qualitative computing: Using software for qualitative data analysis.* Aldershot: Ashgate.

Forster, N. (1994). The analysis of company documentation. In C. Cassell, & G. Symon (Eds.), (1994). *Qualitative methods in organizational research: A practical guide,* pp.147–166. London: Sage.

Gell-Mann, M. (1995). *The quark and the jaguar: Adventures in the simple and the complex.* London: Abacus.

Giddens, A. (1984). *The constitution of society: Outline of the theory of structure.* Cambridge: Polity Press.

Glaser, B. (1978). *Theoretical Sensitivity.* Mill Valley, CA: Sociological Press.

Glaser, B., & Strauss, A. (1967). *The discovery of grounded theory: Strategies for qualitative research.* Chicago: Aldine.

Glaser, B. G. (1992). *Emergence v Forcing: Basics of grounded theory analysis.* Mill Valley, CA: Sociology Press.

Gleick, J. (1994). *Richard Feynman and modern physics. London: Abacus.*

Hammersley, M. (1989). The dilemma of qualitative method: Hebert Blumer and the Chicago tradition. London: Routledge.

Harnad, S. (1987). Category induction and representation. In S. Harnad (Ed.), *Categorical perception: The groundwork of cognition.* Cambridge: Cambridge University Press.

Hindess, B. (1977). *The philosophy and methodology in the social sciences.* Hassocks, UK: Harvester.

Irurita, V. F. (1996). Hidden dimensions revealed—Progressive grounded theory study of quality care in the hospital. *Qualitative Health Research, 6*(3) 331–349.

Isaac, L. W., Carlson, S. M., & Mathis, M. P. (1994). Quality of quantity in comprative/historical analysis: Temporally changing wage labor regimes in the United States and Sweden. In T. Janoski, & A. M. Hicks (Eds.), *The comparative political economy of the welfare state.* Cambridge: Cambridge University Press.

Janoski, T., & Hicks, A.M. (1994). *The comparative political economy of the welfare state.* Cambridge: Cambridge University Press.

Keddy, B., Sims, S. L., & Stern, P. N. (1996). Grounded theory as feminist research methodology. *Journal Of Advanced Nursing, 23* (3), 448–453.

Keil, F. C., & Kelly, M. H. (1987). Development changes in category structure. In S. Harnad (Ed.), *Categorical perception: The groundwork of cognition.* Cambridge: Cambridge University Press.

Kools, S., McCarthy, M., Durham, R., & Robrecht, L. (1996). Dimensional analysis—broadening the conception of grounded theory. *Qualitative Health Research, 6*(3), 312–330.

Kosko, B. (1994). *Fuzzy thinking.* London: Flamingo.

Kumar, K. (1994). The evolution of society: A Darwinian approach. In C. M. Hann (Ed.), *When history accelerates: Essays on rapid social change, complexity, and creativity.* London : Athlone Press.

Lakoff, G. (1987). *Women, fire and dangerous things: What categories teach about the human mind.* Chicago: Chicago University Press.

Lee, R. M., & Fielding, N. (1996). Qualitative Data Analysis: Representations of a Technology: A Comment on Coffey, Holbrook and Atkinson. In *Sociological Research Online, 1*(4).

Lincoln, Y. S., & Denzin, N. K. (1994). The fifth moment. In N. K. Denzin, & Y. S. Lincoln (Eds.), *Handbook of Qualitative Research.* London: Sage.

Lonkila, M. (1995). Grounded theory as an emerging paradigm for computer-assisted qualitative data analysis. In U. Kelle (Ed.), *Computer-aided qualitative data analysis.* London: Sage.

Marsh, C. (1988). *Exploring data: An introduction to data analysis for social scientists.* Cambridge: Polity Press.

Mar, C., & Rossman, G. (1989). *Designing qualitative research.* London: Sage.

Mason, J. (1996). *Qualitative researching.* London: Sage.

Massaro, D. W. (1987). Categorical partition: A fuzzy-logical model of categorization behavior. In S. Harnad (Ed.), *Categorical perception: The groundwork of cognition.* Cambridge: Cambridge University Press.

McNeill, D., & Freiberger, P. (1994). *Fuzzy Logic: The revolutionary computer technology that is changing our world,* New York: Touchstone.

Medin, D. L., & Barsalou, L. W. (1987). "Categorization processes and categorical perception" In S. Harnad (Ed.), *Categorical perception: The groundwork of cognition.* Cambridge: Cambridge University Press.

Melia, K.M. (1996). Rediscovering Glaser. *Qualitative Health Research, 6*(3), 368–378.

Olshansky, E. F. (1996). Theoretical issues in building a grounded theory—Application of an example of a program of research on infertility. *Qualitative Health Research, 6*(3), 394–405.

Ragin, C. C. (1987). *The comparative method: Moving beyond qualitative and quantitative strategies.* Berkeley: University of California Press.

Ragin, C. C. (1995). Using qualitative comparative analysis to study configurations. In U. Kelle (Ed.), *Computer-aided qualitative data analysis.* London: Sage.

Richards, T. J., & Richards, L. (1994). Using computers in qualitative research. In N. K. Denzin, & Y. S. Lincoln (Eds.), *Handbook of qualitative research.* London: Sage.

Robrecht, L. C. (1995). Grounded theory—Evolving methods. *Qualitative Health Research, 5*(2), 169–177.

Rosch, E. (1978). Principles of categorization. In E. Rosch, & B. B. Lloyd (Eds.), *Cognition and Categorization.* Hillsdale, NJ: Erlbaum.

Rucker, R. (1987). *Mind tools: The mathematics of information.* London: Penguin.

Sayer, A. (1992). *Method in social science: A realist approach, (2nd ed.).* London: Routledge.

Seidel, J., & Kelle, U. (1995). Different functions of coding in the analysis of textual data. In U. Kelle (Ed.), *Computer-aided qualitative data analysis.* London: Sage.

Silverman, D. (1993). *Interpreting qualitative data: Methods for analyzing talk, text and interaction.* London: Sage.

Stake, R. E. (1994). Case studies. In N. K. Denzin, & Y. S. Lincoln (Eds.), *Handbook of qualitative research*. London: Sage.

Stones, R. (1996). *Sociological reasoning: Towards a post-modern sociology*. Basingstoke, UK: Macmillan.

Strauss, A. (1987). *Qualitative analysis for social scientists*. Cambridge: Cambridge University Press.

Strauss, A., & Corbin, J. (1990). *Basics of qualitative research: Grounded theory procedures and techniques*. London: Sage.

Strauss, A. & Corbin, J. (1994). Grounded theory methodology: an overview. In N. K. Denzin, & Y. S. Lincoln (Eds.), *Handbook of qualitative research*. London: Sage.

Strauss, A., & Corbin, J. (Eds.) (1997). *Grounded theory in practice*. London: Sage.

Sylvan, D., & Glassner, B. (1985). *A rationalist methodology for the social sciences*. Oxford: Basil Blackwell.

Tudge, C. (1996). *The day before yesterday: Five million years of human history*. London: Pimlico.

Usui, C. (1994). Welfare state development in a world system context: event history analysis of first social insurance legislation among 60 countries, 1880–1960. In T. Janoski, & A. M. Hicks (Eds.), *The comparative political economy of the welfare state*. Cambridge: Cambridge University Press.

Van Maanen, J. (1982). *Varieties of qualitative research*. London: Sage.

Waldrop, M. M. (1994). *Complexity: The emerging science at the edge of order and chaos*. London: Penguin.

Wallerstein, I. (1991). *Unthinking social science: The limits of ninteenth-centry paradigms*. Cambridge: Polity Press.

Walton, J. (1992). Making the theoretical case. In C. C. Ragin, & H. S. Becker (Eds.), *What is a case?* Cambridge: Cambridge University Press.

Wilson, H. S. & Hutchinson, S. A. (1996). Methodological mistakes in grounded theory. *Nursing Research, 45*(2), 122–124.

SUBJECT INDEX

ISBN 0-12-214640-9

DATE DUE
